"To Telena, who inspired this book, and Charley who will add the essential details I missed in about 20 years. Fortunately, the universe is patient."

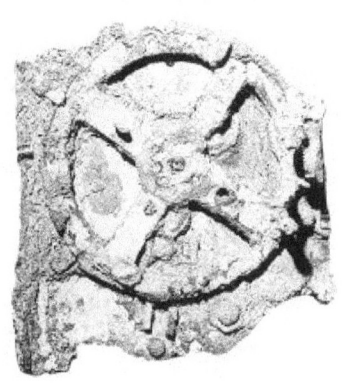

ZIGAMUS GLORIBUS PUBLISHING COMPANY
4920 S. LOOP 289 – SUITE 101
LUBBOCK, TEXAS 79414

TABLE OF CONTENTS

PREFACE   4

CHAPTER 1 – BEGINNINGS   17

CHAPTER 2 – THE GAME OF LIFE   23

CHAPTER 3 – TEILHARD AND HIS ARROW   34

CHAPTER 4 – OF WATCHMAKERS & DESIGN   43

CHAPTER 5 – COMPLEXITY & THE EVOLUTION OF MAN   62

CHAPTER 6 – COMPLEXITY & THE COSMOLOGICAL RECORD   73

CHAPTER 7 – COMPLEXITY & THE ATOMIC RECORD   78

CHAPTER 8 – TEILHARD'S ARROW – YOU ARE HERE   82

CHAPTER 9 – THE FACE OF GOD   90

CHAPTER 10 – INTELLIGENT PROCESS   96

CHAPTER 11 – DEPARTURE FROM THIS STRANGE WORLD   100

CHAPTER 12 – EMERGENT SPECULATIONS   108

CHAPTER 13 – EMERGENT BORDER PHENOMENA & THE QUANTUM/RELATIVITY RIDDLE   **116**

CHAPTER 14 – CONSCIOUSNESS ALONG TEILHARD'S ARROW   **141**

CHAPTER 15- CONSCIOUSNESS AND ARTIFICIAL INTELLIGENCE   **158**

CHAPTER 16 – GENESIS 1.0   **172**

CHAPTER 17 – THE ANATOMY OF EVIL   **181**

CHAPTER 18 – INSIDE HEAVEN'S GATE   **213**

CHAPTER 19 – POSTSCRIPT   **218**

END NOTES   **220**

PREFACE

GEORGETOWN UNIVERSITY CAMPUS

I have enjoyed solving complex intellectual puzzles for most of my adult life. One of the greatest challenges has been to assemble a so-called "Theory of Everything" to explain the emergence of human consciousness and whether the existence of God can be scientifically validated.

Of course, this is a daunting task and even Einstein was not up to the challenge. I certainly do not consider myself to occupy such an elevated station. That said, I have been obsessed with this topic since my junior year at Georgetown University in 1975. It will be 43 years as of the date of this writing. The questions pursued in this book were sparked by a theology class I attended in the fall of 1975 presented by Thomas M. King, S.J.[i]

Of course, over this span of time, there have been other matters to consider, family issues to attend to, and the twists and turns of everyday life which seem so important at the time. All in all, it is somewhat of a miracle that one topic has occupied such a commanding position in my consciousness at all.

THOMAS M. KING, S.J.

Father King dedicated his life to the study of Pierre Teilhard de Chardin, a Jesuit priest and paleontologist who grappled with finding a common denominator to embrace an abiding love of science and faith in Jesus Christ.[ii] As a paleontologist, Teilhard was fascinated by the concept of evolution as advanced by Charles Darwin and contemplated whether this theory could somehow be reconciled with Christian theology. Teilhard was not an armchair philosopher, but actually ventured forth to the hinterlands of Africa and other environs in search of the prehistoric precursors of modern man and is credited with numerous discoveries which advanced the search for our human origins. Such scientific exploration and curiosity, however, came with a heavy price as his efforts were not appreciated by his colleagues in the Catholic Church and the Vatican itself. Reminiscent of the stigma visited upon such luminaries as Galileo, members of the Church community were threatened by the questions posed by Teilhard and whether the possible answer would lead to the elimination of God as the guiding force and creator of the universe. Conversely, the scientific community decried Teilhard's theory as tainted by religion and mysticism, the antithesis of the science. This "fault line" between science and theology still exists today and is seemingly more pronounced than ever.

What I remember most about Father King was his incredible intellect, curiosity, and – above all, his kindness. Like Teilhard, he sought to reconcile his Christian theology with the enormous advancements in scientific research in all areas spanning evolution to physics. No topic was off limits to Father King and, alas, as the singular Jewish student in that Georgetown class of '75, I was the leading skeptic. I thought my job was to accentuate that "fault line" between science and theology. I questioned whether the Christian concept of God or any derivation thereof had any place whatsoever in intelligent discourse about the origin of man and consciousness. It seemed like a lot of hocus pocus to me.

Notwithstanding the above, Father King's class on Teilhard changed the intellectual and spiritual focus of my life. A spark ignited my curiosity as to whether existing scientific theory as enunciated by Charles Darwin truly explained the panoply of life. Yes, I could understand that the principle of "natural selection" characterized Darwin's Theory of Evolution and that a tree of life with many branches could be traced to humble origins – whether the prehistoric precursor to the modern horse or man. However, the "randomness" of evolution troubled me as it seemed to ignore the existence of a decided direction in the process itself. Teilhard (via Father King) introduced the missing link in my early ruminations that there might be an overriding pattern or direction to evolution itself – complexification. Teilhard's theory supplanted classic evolution by asserting that the existence of reflective human thought was not simply a byproduct of objective laws of nature, but was the personification and culmination of the "Phenomenon of Man"[iii]– the title of his seminal work in this area. Teilhard introduced the concept that evolution was not random, but evidenced a verifiable and objective physical progression of matter from its simplest to more complex forms. Evolution was not a tree with many branches, but akin to a ladder representing increased stages of complexity from simple cells to organisms of greater complexity. Evolution did not disturb or threaten Teilhard – he embraced the continuum of simple structures arising in the primal currents of an ancient sea and advancing rapidly in complex progression from fish to amphibian to reptile to mammal to primate and to man. With the advent of the human brain, matter was able to "reflect upon itself" for the first time, and according to Teilhard, that was truly a revolution.

Teilhard further theorized that a preexisting consciousness which he termed "Omega" or the "Omega Point" was like a gravitational force which pulled matter to increasing levels of complexity – very much like a spiritual magnet of sorts.

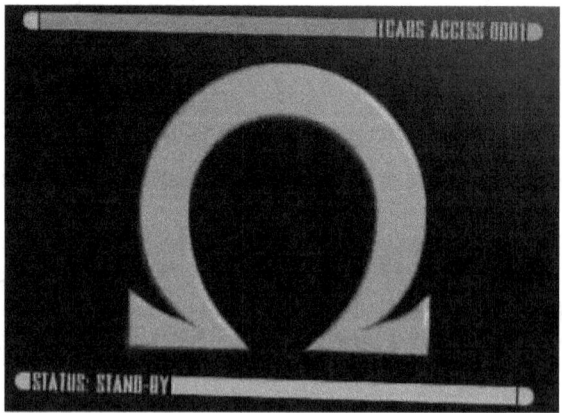

THE OMEGA POINT

No surprise that Teilhard attributed the Omega Point to Christ Jesus which is discussed in some length in the final chapter of the Phenomenon of Man.[iv] Such is the scientific and theological bridge spanning the otherwise vast chasm between science and theology – evolution and spirituality.

In the years following this introduction to Teilhard and Father King, I continued the process of ruminating whether there was a direction to evolution or whether – in the words of Richard Dawkins – we were simply the byproduct of random natural forces a/k/a the "blind watchmaker".[v] Of course, some of this inquiry was directed to an increasing sense of my own mortality and whether all of this religious dogma had any import whatsoever other than being a convenient crutch or escapist delusion.

I also keenly tracked other developments in science and physics which intrigued me. One area known as "Emergent Systems Theory" examines unique and unpredictable behavior which emerge within a larger system. Examples include the flocking behavior of birds, the origination of storms system such as tornadoes and hurricanes, and even the sudden rise or crash of the stock market.[vi] The concept of "emergence" became a buzz word of sorts indicating that the "whole was greater than the sum of its parts".

EMERGENT SYSTEM – FLOCKING OF BIRDS

One of the first efforts in this area was spearheaded by mathematician John Conway who created a model known as the "Game of Life". Before the advent of computers, Conway created a game played by his college students which set forth simple rules for basic structures known as automata or simple cells. Simple rules acted out by the participants replicated evolutionary forces by determining whether single automata would survive into the next generation or die out. Each progression in the game was considered a generation and the goal was to play out as many generations as possible before the college kids graduated. And in so doing, something unpredictable occurred - *the automata became more complex over time*. This happened *every time* and the simple rules of the "Game of Life" could not explain the phenomena. With the advent of computers, the Game of Life was replicated in a digital world over millions of generations (sans the kids) and what was uncovered was nothing less than amazing. With each succeeding generation, the automata became more complex and exhibited characteristics which were not predicted by the simple rules at inception. Incredibly, complex mathematical structures with the identical geometric forms emerged game after game which were assigned names such as pulsars, sliders, toads, and beehives.

The electronic structures which emerged in these pre-computer simulations are now referred to as "electronic life" or "e-life" which has developed into its own subfield.[vii] To this day, no scientific explanation has been proposed as to how and why these structures emerge. However, there is universal agreement that the phenomena observed evidences a complexification over time.

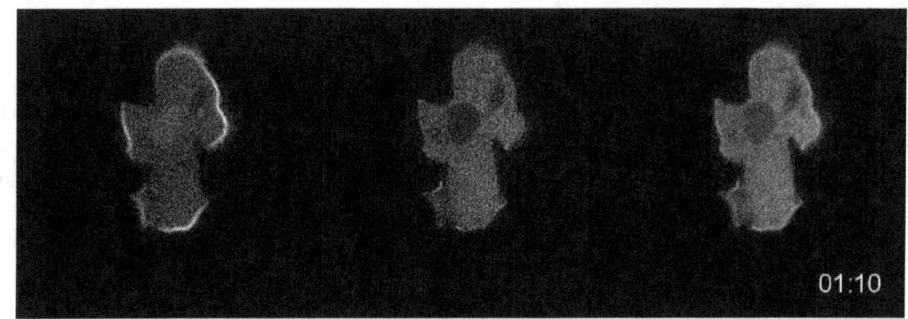

E-LIFE PROGRESSION OVER TIME

From my perspective, e-life and its analogues in cosmology, biology, physics, and paleontology set forth a principle of nature akin to "gravity" which underlies the evolutionary processes of non-organic and organic structures. This principle as first intuited and enunciated by Teilhard may explain not only the evolution of man but the universe itself and our place in it.

All of this brings us back to the title of this work – *Teilhard's Arrow*. I have always tried to express abstractions in some geometric or artistic form to better understand the principles involved. Hence, the metaphor of a seemingly simple arrow to illustrate a most complex proposition. The arrow embodies Teilhard's central theological and scientific view that evolution is proceeding in the direction of greater complexity. And so, it is with Teilhard and my ruminations as to whether there is a direction in the progression of evolution and consciousness itself. Analysis of the paleontological, biological, atomic, and cosmological record reveals the distinct presence of an arrow evidencing Teilhard's theory of complexification over time and in all disciplines. Whether one views the complexification of cosmological structures from the emergence of the so-called Big Bang to present, the complexification of the fossil record from simple one cell structures to dinosaurs and mammals, to the complexification of subatomic particles such as gluons to quarks to protons to atoms and molecules and matter – and finally the complexification of electronic life – unmistakably something remarkable is afoot. Shall we consider this phenomenon to be entirely the random process of objective grinding rules which drive evolution – or is there an overriding point of physical and spiritual gravity which is ever drawing matter to higher levels of complexity – the Omega Point as envisioned by Teilhard?

As will be discussed, a verifiable and objective "arrow" extends from simple origins to greater levels of complexity in all things and disciplines. This is the spine of complexification theorized by Teilhard. The fascinating question to be explored is whether this arrow is just a random process or the fossil manifestation of a forward direction in time, complexity, and consciousness? Is evolution simply an objective – mindless – grinding process of random phenomena?

The end point of my ruminations dictates that the answer is decidedly "No". There is – in fact – an arrow pointing toward a glowing, super-human – and *contemporaneous* consciousness very much as Teilhard postulated but with critical distinctions as outlined herein.

As will be explored in this essay, there is a direction as evidenced by Teilhard's Arrow. We can uncover its presence in nature and physics and intuit is *co-existence with us* going forward in time. The core conclusions of this work are as follows:

1. Complexification is a force of nature which will one day be reduced to formula with the precision of Newton's gravity.
2. Evolution is propelled not by natural selection but by complexification.
3. Mankind and the emergence of thought are not the culmination and pinnacle of evolution but simply a rung along the ladder of complexification which has likely *passed us by eons ago*. We are simply a point along the continuum of complexity, probably somewhere in the middle.

4. Complexity dictates that structures (organic and otherwise) exist contemporaneously with us which are as far advanced from us as we are from an amoeba.

5. The complex structures which exist beyond our level of awareness have a super consciousness with a defined center much like Teilhard's Omega Point. The arrow pointing toward Omega can also be extended backward in time with decreased levels of complexity and is detected in all of our scientific disciplines.

6. The super-human Omega Point would be incomprehensible to humankind at its lower level of complexity and consciousness. Point of fact is that Omega would appear to us as a God. It would meet every definition of God. It would be miraculous as it would be inexplicable and incomprehensible within the limited bounds of our intellect.

7. As Teilhard opined, this Omega would manifest itself in the universe in everything we do. We could be aware of its presence. It would hear our prayers.

8. In the same manner that logic and science can be utilized to demonstrate the existence of "God", so also would the existence of an evil sentient force be established. The systemic forces of entropy

and synthesis find their manifestation in every emergent level spanning the atomic to the inorganic to the organic. Within the context of human thought these basic countervailing forces can be construed as representing good and evil.

Teilhard's arrow points decidedly below and above us. Once this structure is scientifically established, a permanent bridge will be constructed between science and theology. Such development will shatter existing paradigms of who we are and our role in the universe.

And finally, a word of caution for my faithful readers. It is best to consider this work as constituting two parts: Part I (Chapters 1-10) sets out my theory which relies on objective scientific data in fields ranging from cosmology to physics to biology, to make the case for what this writer considers to be the existence of "intelligent process'. Part II (Chapters 10-18) departs from a strictly scientific perspective and engages in an imaginative and yes – speculative journey – into the far reaches of physics and cosmology in an effort to formulate a comprehensive "Theory of Everything." To my critics, all I can say is – have at it! At the very least, I have advanced some novel concepts which will suggest new approaches to pursue.

CHAPTER ONE

BEGINNINGS

THEOLOGICAL AND COSMOLOGICAL BEGINNINGS

I have always been fascinated by science. As a child, my curiosity about the world around us and universe was unbounded. It still is. However, I have come to believe that the underlying antagonism between science and theological thought as espoused by most scientists in the field is as uncompromising and rigid as the Catholic's church's condemnation of Galileo. Notwithstanding his house arrest and very real threat of torture and death, on his death bed Galileo was said to still utter to the attending priest – "and yet it moves…"[viii]

And to the greatest scientists and geneticists and physicists of this day and age, I will proudly utter my 21st Century rejoinder – "A higher power, whether God – Christ Omega – Yahweh, or whatever nomenclature we hominids choose to describe the unfathomable, still reigns supreme.

The discussion on the subject as to whether the emergence of mankind and intelligence is the product of automatic evolutionary principles or the "intelligent design" of a preexisting creator has sparked enormous debate. Unfortunately, the proponents of either theory have assumed intractable positions which discredit the underlying logic of both extremes. The evolutionists adhere stubbornly to the 19th Century observations – scientific conclusions – and – yes – conjectures of Charles Darwin and his seminal work, the Origin of the Species. As will be discussed in Chapter 4, the underlying theoretical assumptions of Darwin which seemed logical in the pre-genetic world of his century and the next half century thereafter, conflict with contemporary biology, genetics, and probability theory. Natural Selection – the driving force of Darwin's evolutionary theory – cannot as a matter of "science" explain the complexification of life on Earth. Natural selection which is premised on the survival of the fittest can explain simplistic changes in the structure or color of a butterfly, but not the complexification of life which underlies the evolution of simple life forms to human intelligence. Such occurred "Not By Chance!"[ix]

Conversely, the so-called theologians on the other side are no less adamant and blindsided in their opposition to Darwin and the "evolutionists". The "Creationist" school takes a literal interpretation of the bible and has enormous difficulty reconciling the Big Bang with Genesis – let alone the appearance of dinosaurs. But for the Church's final apologia that Galileo was right, the Church itself might still maintain that Earth is the center of the solar system and universe itself.

That said, the term "intelligent design" which has generated such acrimony between scientists and theologians – does suggest an "intelligent" middle ground which best explains the progression of complexity and the forward direct of evolution itself. However, this middle ground cannot be achieved without modifying the term from the oft used "intelligent *design*" to "intelligent *process*" as suggested in this work. And that makes a universe of difference.

Intelligent Design is based on the notion that a preexisting God Creator intentionally "designed" the universe and mankind's place therein. The underpinning of this theory of a "First Cause" dates back to the writings of Thomas Aquinas. In a more modern context, William Paley in his 1802 work Natural Theology advanced the analogy of a watchmaker.[x] In the same way the complex machinations of the watch were predesigned and executed by the watchmaker so is the panoply of life designed by a universal creator and God force which preexisted all creation. Intelligent Design is the post-modern expression of this notion that the complexification of life must have been preceded and designed by God. Of course, the difficulty of Intelligent Design is that it cannot be scientifically proven. Distilled to its essence, it demands that its proponents maintain the rigidity of Galileo's enemies.

There is a middle ground which has its roots in the explorations and writing of the early to middle 20th Century theologian, Teilhard de Chardin. Teilhard was educated as a Jesuit and yet was fascinated by the concept of evolution and its manifestation in the field of paleontology. This Catholic Priest ventured forth to the hinterlands of Africa in search of the hominid predecessors of mankind and was credited with important discoveries in that area.

Teilhard sought to combine his abiding belief in Christ with the undeniable evidence of evolution. The theory he proposed as espoused in his seminal work, *The Phenomenon of Man*, set forth a paradigm which for the first time, spanned the intractable position of the creationists and evolutionists. In Teilhard's view, the *Phenomenon of Man* was propelled by a divine force of internal centricity which moved all life on earth to higher levels of complexity. This "spirit" of matter – was drawn to high level of complexity by a force akin to a spiritual magnetism or gravity – Omega. The Omega Force – or the ultimate complexification of spiritual life was the Christ Force which in Teilhard's theory – explained the upward complexification of matter from its simple atomic structure to the burgeoning of life in a prehistoric pond – to complex multicellular forms – to mankind itself.[xi] To Teilhard's enormous credit, the notion of what we now refer to as "intelligent design" was combined with evolution to form a workable theory. That said, the theory is still subject to the scientific infirmity as to the inability to prove up the existence of a "first cause". Teilhard is light years ahead of the so-called modern ramblings of the intelligent design proponents, but still misses the mark. Where Teilhard is correct, is the starting point of this essay.

The complexification of life demonstrates the existence of a process referred to as evolution. However, unlike the proponents of Darwin, the evolutionary process is not mindless and undirected. And... unlike the meanderings of the creationists and intelligent designers, there cannot be a preexisting first cause or creator which designed all of creation retroactively. However, the process of evolution and its forward motion towards complexity has created an intelligence which *coexists* with mankind's existence circa 2017.

Mankind is not the pinnacle of evolution but simply an intermediate form somewhere near the middle. As I peer at the marine fish in my aquarium and they "peer back" with some glimmer of intelligence, so to, a higher form of intelligence *at this moment* is aware of my presence. As I might be viewed God-like to my pet and overfed Oscar nearing the surface for its morning meal – so would the entity of greater complexity be viewed from my perspective. It would be all knowing – probably eternal – and "God" would be an excellent description. However, the presence of this higher power or "Omega" would have resulted from the natural force of complexification thereby rendering its intelligence not as preexisting but the byproduct of a process – thus the term "intelligent process". In a literal sense, this intelligence would preexist any of us alive today, but it would have arisen as a natural process of higher complexification in the continuum of development. A better way of understanding this concept is that the Omega force would represent a higher dimension of reality and cognition which coexists with mankind but is otherwise undetectable. In that sense, it would appear to be a miracle.

But just like complexity itself, the author is getting ahead of himself. Let's start out by looking no further than our computer.

CHAPTER TWO

THE GAME OF LIFE

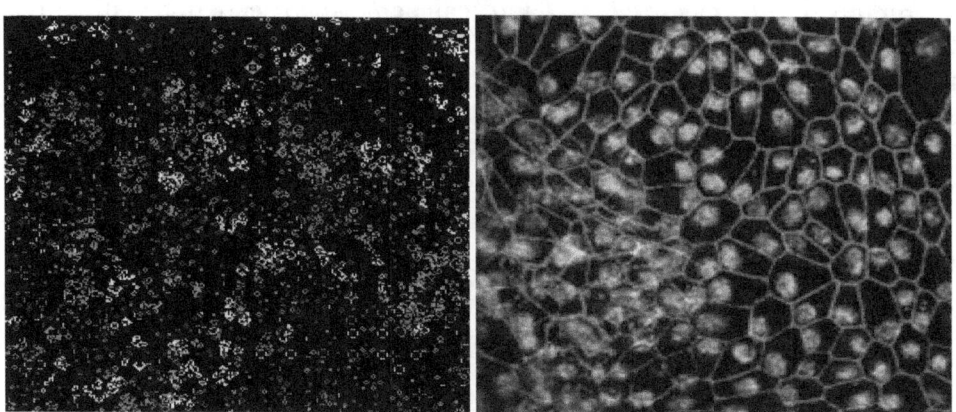

ELECTRONIC & ORGANIC LIFE

There something inexplicably grand going on within the guts of your PC. Google the term "Conway's Game of Life Online" and a plethora of links to what appears to be a simple video game appears. The rules proceed quite simply at first. What follows is nothing less than magic without the gimmick. Houdini himself could not expose the sleight of hand.

The game begins with an electronic checkerboard and two game pieces – black and white squares.[xii] Given that the name of this simulation is the "Game of Life", it is befitting that we refer to each game piece as a "cell" and assign two possible positions to each – "alive" or "dead", respectively – black or white. Each cell interacts with its adjoining eight cells which are horizontally, vertically, or diagonally adjacent. In this two-dimensional electronic world – only four rules apply – nothing more complex than that. Inhabitants of this "electronic flatland" can achieve no academic distinction greater than the ability to recount but four basic rules which dictate the interaction of all things black and white – horizontal, vertical, and diagonal.

1. Any live cell with fewer than two live neighbors dies, as if caused by over-population.
2. Any live cell with two or three live neighbors lives on to the next generation.
3. Any live cell with more than three live neighbors dies, as if by overcrowding.
4. Any dead cell with exactly three live neighbors becomes a live cell as if by reproduction.

Below is a simple illustration of how the rules apply in our Electronic Flatland.

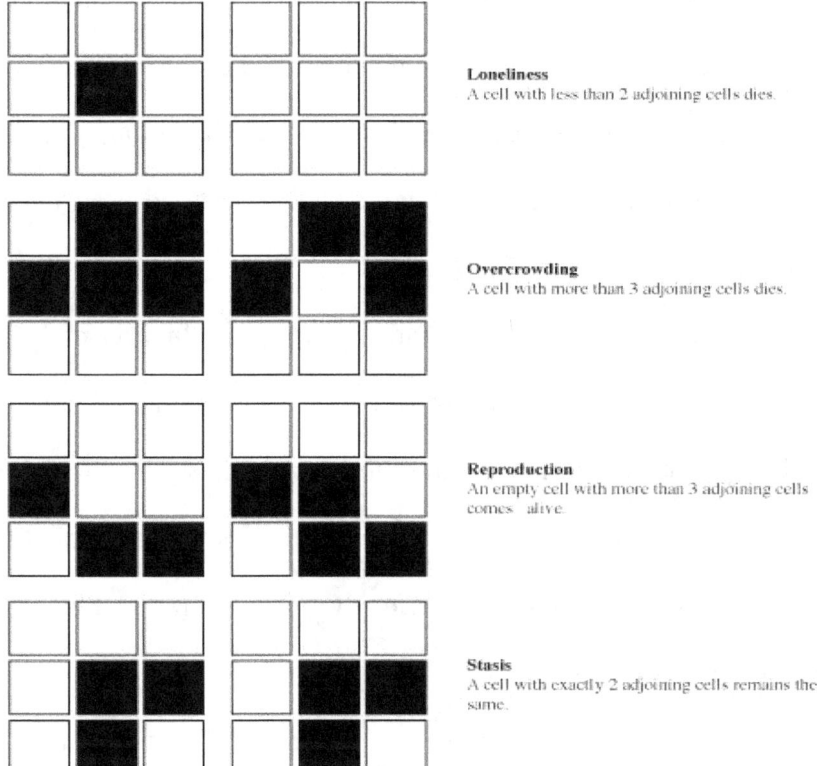

Loneliness
A cell with less than 2 adjoining cells dies.

Overcrowding
A cell with more than 3 adjoining cells dies.

Reproduction
An empty cell with more than 3 adjoining cells comes alive.

Stasis
A cell with exactly 2 adjoining cells remains the same.

The "game of life" was devised by British mathematician John Horton Conway in 1970. Conway was intrigued by the research of mathematician John Von Neumann in the 1940's who attempted to divine a machine which could hypotheically reproduce an exact copy of itself. In an age before the advent of home computers, Conway invented a checker board game where student players employed the above four rules proceeding from page to page, one genration at a time. In the first iterations of this game which proceeded in slow student-powered motion, Conway identified unique configurations of blocks which seemed to appear out of nowhere and evolve over time. He was astounded, to say the least. These early black and white images penciled in by students and faculty were given names which are still used today – gliders and blinkers and beehives, and so on.

With the advent of computers and the mass production of PC's – researchers and amateurs alike transformed Conway's Game of Life from mechanical to electronic – and the results were fascinating. The first images that Conway "discovered" now dubbed as e-organsims or electronic life – were replicated and surpassed. A previously undiscovered e-universe as inexplicable then as it is today emerged.

The game was played in a series of "generations" in which the four rules are applied to every cell in the initial pattern. In generation-1, the four rules are mechanically applied endowing some cells with the gift of life, death, reproduction, or stasis. Given the simplicity of this process, your PC can complete the calculations of this first generation in the blink of an electronic eye - .00001/second. The next configuration of the cells at generation-2 is slightly different than its precedecessor and so on. Tack on just five or ten generations and you are in for a kaleiodscopic treat.

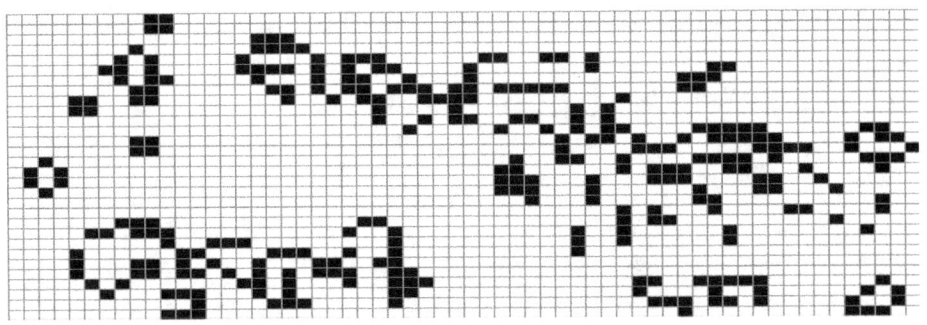

E-ORGANISMS ABOUND IN GAME OF LIFE

To the casual observer of electronic cells – or automata as they are referred to in a facinating discipline known as e-life or electornic life, it just looks like the random moves of a few game pieces – nothing more. However, increase the rapidity of the generation progression to the maximum speed of your PC – and something quite remarkable occurs. Patterns seem to appear out of nowhere that have discernable characteristics as they replicate over and over again in an electronic zoo of sorts. Some of these patterns spin around like a top, or blink incesstantly like a traffic light, or "glide" across the screen in meandering patterns of four cells. The very fact that these patterns seemingly appear out of nowhere and can be observed by artificial life electronic officionados worldwide is without embellishment – *astonishing*. One transitions from *astonishing* to *miraculous* when it is readily admitted that no one on the planet has the foggiest idea or explanation as to how or why these patterns or electronic organisms come to be. In the identical fashion a zoologist would identify different species of life – our e-zoologists have categorized the e-life species thus:

3P2.1 Blinker	4.1 Block	4.2 Tub	5.1 Boat	5P4H1Y1.1 Glider	6.2 Ship
6.4 Beehive	6.5 Barge	6P2.1 Toad	6P2.2 Beacon	7.2 Long Boat	7.4 Loaf
8.7 Pond	8.8 Mango	8.9 Long Barge	12.41 Half-Fleet	14.533 Half-Bakery	

THE MULTIPLICITY OF E-SPECIES IN GAME OF LIFE

Since the "discovery" of the first series of e-life creatures, the above list has expanded to such e-species as boats, beehives, loafs, eaters, ponds, spaceships, bipolar, tripoles, and … the list goes on! Notably, I employ the word "discovery" as the identification of these e-creatures is no different than the discovery of a new species of fish or reptile. These e-creatures have distinct characteristics and can be objectively verified by multiple observers over time. They exist in their own electronic world in a reality that is eerily reminiscent of our own.

Into this e-petrie dish of sorts, enter one of the most fascinating e-creature of all – the "R Pentomino". This e-creature looks a bit like a lower case "r" – but don't let the simplicity of its basic design fool you. The R Pentomino is inherently unstable and "mutates" into new e-creatures at an astounding pace.

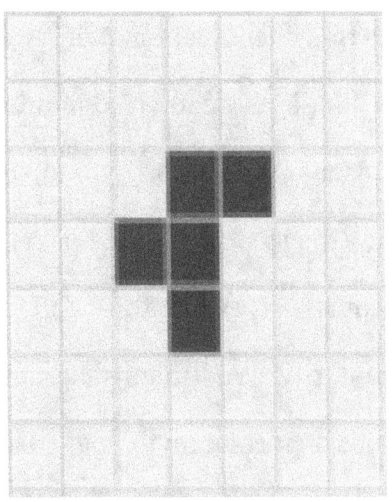

R PENETOMINO

At generation 1, above, the R Penetomino looks a bit like a periscope. At generation 2, it mutates into a beehive and continues this progression becoming increasingly unstable over time. By generation 9, it has broken into several pieces and continues on kaleidoscopically until it appears to assume a stable pattern at generation 48. Of course, this is short lived, and new e-objects are ejected about twenty generations later which proceed along their own evolutionary pattern. A plethora of fascinating e-organisms emerge resembling flowers, traffic lights, and ships. These e-objects then combine with each other bringing forth more diverse patterns which appear to interact with each other and to assume defined roles. Fast forward thousands of generations on any PC and millions of generations on more powerful computers, and this phenomenon of development continues on.

In 2010, Andrew J. Wade, a Canadian computer scientist, discovered a self-reproducing e-organism at generation 33,699,586! Very much like the function of RNA within the micro world of organic cells, our e-organisms can now reproduce exact copies of each other and continue on along an evolutionary wave of complexity. Wade named his new "discovery" – Gemini – after the Greek twin Gods of ancient mythology. This description seems to fit as the structure of Gemini forms two seemingly identical structures which proceed in development parallel to each other.

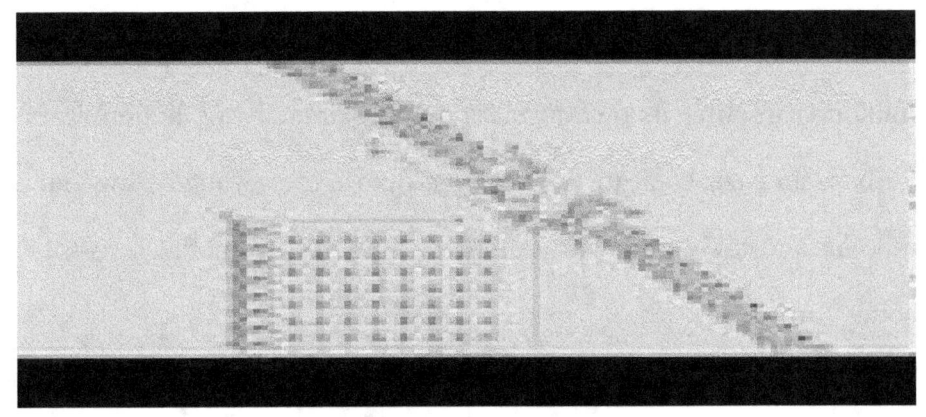

GEMINI

To add to this fascinating discovery – Wade noticed that a thin line of patterns emanated from one e-twin to the other and proceeded back and forth as if exchanging information! Further observation confirmed that these tiny objects were indeed - *gliders* – the early e-organisms first discovered with the ability to move in concerted patterns across the computer screen. To the amazement of all, the e-gliders were somehow transmitting information from one twin to the other culminating in an exact reproduction of a new e-organism! To say this is mysterious is an understatement.

The truly fascinating question is how far can this evolution continue and what might emerge within the confines of a super computer at say – generation 10 trillion. Might we see electronic single cell organisms mutating into multi-cellular creatures and so on. What exactly might emerge in this undulating e-broth reminiscent of the ancient oceans of our own planet 700 million years after its own formation approximately 4.5 billion years ago? Perhaps an e-sea creature or even an e-dinosaur crawling its way out of the e-ocean and taking the first tentative steps from your flat screen to your coffee cup?

So, what precisely is going on here? To be sure, decades after the first discovery of this e-phenomena, nobody knows for sure. However, two intriguing conclusions jump out at the observer – (1) the patterns which emerge from one generation to the next are *not* "random", and (2) the patterns take on more complex forms and patterns over time. These two conclusions literally form the spine of the thesis of this work – both figuratively and literally. It is an arrow which extends from the elementary to complex. It can be observed in the electronic world of Conway's Game of Life and in every area of science and nature devised by mankind. As will be discussed, this "arrow" underlies the structure of biology, paleontology, physics, astronomy, and e-life. From initiation, some underlying force or process is driving basic elementary units to greater levels of structure and development over time. Whether one observes the origins of matter in the first micro seconds after the Big Gang or the paleontological records of the first pre-humans to modern humankind –an arrow pointing to ever increasing levels of complexity emerges. This is not conjecture. It is as plain as day. This arrow is as apparent as the shape of metallic patterns which encircle a bar magnet.

Superimpose the progression of moving parts in any discipline from elementary physics to biology to astronomy and e-life – and a pattern reminiscent of metal on a bar magnet emerges. This pattern is pointing ever upward toward greater complexity. Stand back a few feet or a few miles – a few light years – or billions of years and only one conclusion is possible. The image has a clear shape – and it is an arrow – pointing ever upward.

CHAPTER THREE

TEILHARD AND HIS ARROW

COMET IMAGE COURTESY OF NASA

My first lecture by Father Thomas King at Georgetown University in the Fall of 1975 had a profound effect on my view of the universe and how a particularly inquisitive 21-year-old acclimated thereto. This was a period of intense intellectual growth as my curiosity and libido were at their peak. I remember that year as being as a personal renaissance of sorts. Georgetown was a haven of some of the most influential minds in the areas that mattered most to me – philosophy, science, and theology.

Father King was a phenomenal soul and intellect. He was as brilliant as he was kind and well known for his passionate pursuit of the writing of Pierre Teilhard de Chardin, a relatively obscure Jesuit theologian and scientist. Little did I realize at the time, but Teilhard became my own organizing force in ascertaining my own sense of the universe.

TEILHARD de CHARDIN

Teilhard was born is France in 1881 in the Chateau of Sarcenat at Orcines. His father was an amateur naturalist and encouraged Teilhard to consider some of the great minds of the day, not to mention the writing of Charles Darwin and his seminal and controversial work – Origin of the Species. In tandem with his exposure to the world of science and philosophy, Teilhard was educated at a Jesuit College and began his training in Jesuit society in 1899. As a young Jesuit, Teilhard was exposed to the traditional doctrine of the Catholic Church and its "Christ Centric" view of all creation and the universe. However, what made Teilhard truly unique was his unbridled scientific curiosity in the areas of paleontology and geology which stood in stark contrast to the orthodox ecclesiastical doctrines of the Vatican. Following in the footsteps of Galileo whose works were banned by the Church, Teilhard confronted the identical opposition and ostracism. In 1925, he was ordered by the Vatican to leave his teaching position in France and sign a statement withdrawing his positions on the interconnection between the divinity of Christ and the Theory of Evolution. He faced continued criticism and condemnation by the Church until his death in 1955. Notwithstanding the vitriolic attacks of the Church, Teilhard never abandoned his belief that the Gospel of Jesus Christ could be reconciled with the writings of Charles Darwin. Neither fish nor fowl – he was the epicenter of controversy meted out by the very extremes he sought to unite – the Catholic Church and the great scientific minds of the early 20th Century. The controversy he ignited still reigns supreme today.

It was Teilhard's work as a paleontologist which had the most profound effect on his intellectual and spiritual development. He was a member of the first digging teams assembled by luminaries including Charles Darwin who studied the remains of early man. During this period, he worked at the paleontological laboratory at the Museum National d'Histoire Naturelle in Paris focusing on the mammals of the Tertiary period – 66 Million years ago. The hard-scientific evidence witnessed and uncovered by Teilhard "the scientist" led to his whole-hearted endorsement of the precepts of Darwinian Evolution. Notwithstanding the Bible's declaration that mankind emerged in a perfect garden in a place called Eden, Teilhard ascribed to a hierarchy of development which progressed from the most basic organism to the most complex. In his seminal book, the *Phenomenon of Man*, Teilhard detailed the progression of primate evolution from the first primitive forms such as Australopithecus (5.3 million years ago) to Neanderthal (200,000 years ago) to modern man (Homo Sapiens).

Teilhard's research led him to the inescapable conclusion that evolution defined the progression of not just biology but matter itself from the simple to the complex. His conclusion was as stunning as it was controversial – there was a direction in the progression of evolution as personified by an arrow and its preexisting terminus – a point of all divinity and intelligence and soul called "Omega" or the "Omega Point".[xiii] Omega was the end all of creation and the very embodiment of Christ Jesus to Teilhard. Omega was the extreme and final pole of a "spiritual magnet" which pulled all matter to greater levels and degree of complexity – from the atomic to the organic to the multicellular to the invertebrates to the reptilian to the mammalian to the primates and to man. Teilhard considered this upward driving force to be a principle of nature itself in contrast to an equally powerful and yet opposite force which propelled matter and life in the reverse direction of decay, disintegration, and death – the concept of entropy.[xiv]

Teilhard didn't stop with the emergence of man, but traced this "hominization" as he phrased to ever higher level of complexification[xv]. With mankind, matter achieved its highest level of complexity and gave birth to a spiritual and intellectual light which changed the universe itself – "thought".[xvi] The miraculous ability of mankind to achieve consciousness was in Teilhard's view, matter's evolution from simple particles to the ultimate complexity. When mankind achieved the ability to reflect, matter was able for the first time in the history of the universe to reflect upon itself. Consciousness was the sum total of complexification and hominization as embodied in modern man and all of his musings, machinations, and creations.

With the advent of thought, Teilhard pointed his arrow ever upward as the sidereal force of spiritual energy predicted in greater levels of human complexification to be evidenced in the future by a centrification of thought itself into a "mega consciousness" or "Noosphere" forming a biosphere around the world[xvii] – to the ultimate union with the center of all centers – Christ Jesus or the Omega Point.[xviii]

Central to Teilhard's paradigm is not only the existence of Omega – but its "preexistence" from the beginning of matter itself. In the beginning, there was Omega and like a great spiritual magnet, it pulled all matter to greater degree of centricity culminating in matter's grandest product of all – mankind and thought. In the final chapters of the Phenomenon of Man, Teilhard conjectures as follows:

"Christ and Omega thus predated matter and its divine centricity literally pulled matter up by its bootstraps culminating in mankind as the sum total of this progression."[xix]

It is these two concepts – the placement of mankind at the top of the hierarchical progression of matter and the preexistence of Omega which define Teilhard's philosophy. It is also the point of departure for this writer and the discussion of the arrow set forth in this work. With the greatest respect for Teilhard and Father King – this author rejects these two fundamental principles which underlie Teilhard's work.

The scientific evidence unmistakably points to the existence of an arrow in the direction of higher levels of complexification. The existence of this principle which Teilhard refers to as a sidereal force, is confirmed by the discoveries in the area of physics, biology, astronomy, and e-life. However, this force is little understood at the beginning of this 21st Century although its trappings can be seen everywhere and in all scientific disciplines. That said, there is NO evidence whatsoever that this arrow points to mankind as the ultimate progression. In my view, such a conclusion as advanced by Teilhard in not scientifically based and a false assumption. Indeed, Teilhard, in all of his brilliance, seems to have fallen victim to the false gravity of the Church itself and its grandiose obsession that mankind still reigns supreme as the epitome of God's creation. As will be further explored in this work, a more reasoned view is that mankind exists in tandem with a hierarchy of developing forms in paleontology, physics, astronomy, and e-life. No different. That said, there is a panoply of more primitive forms which preceded the evolution of man and – yes – a hierarchy of forms which hypothetically exist right now which are *more complex than mankind*. We are likely located someplace "in between" in the hierarchical progression of matter. There is likely a level of complexification which co-exists with us at this moment and is of a greater level of centrification and awareness. I say this with a keen awareness that evolution does not destroy earlier forms in the upward progression of its arrow. Single cell organisms co-exist with the most complex configurations such as DNA and the human brain. Likewise, sharks which have remained unaltered by evolution for millions of years co-exist with amphibians, reptiles, mammals, primates, and man.

The intriguing conjecture set forth herein is that Teilhard is correct in his conclusion that an arrow points to greater complexification. He is also correct that a great spiritual energy or Omega co-exists with us at this very moment and should be able to make its presence known. However, I believe he is incorrect that this Omega Point preceded matter and man. Logically, the force of complexification exceeded mankind traversing beyond mere thought and consciousness to a mega thought and mega consciousness and on to Omega which emerged as a byproduct of this yet unnamed force which propels complexification. Therefore, Omega now exists but as a result of "intelligent process", not "intelligent design". Very much like the Oscar fish which stares out at me with limited cognition from the confines of its secure aquarium, so do we stare upward towards some unknowing void which we cannot grasp at our elementary level in the hierarchical progression of life. Such is the subject of further discussion.

CHAPTER FOUR

OF WATCHMAKERS & DESIGN

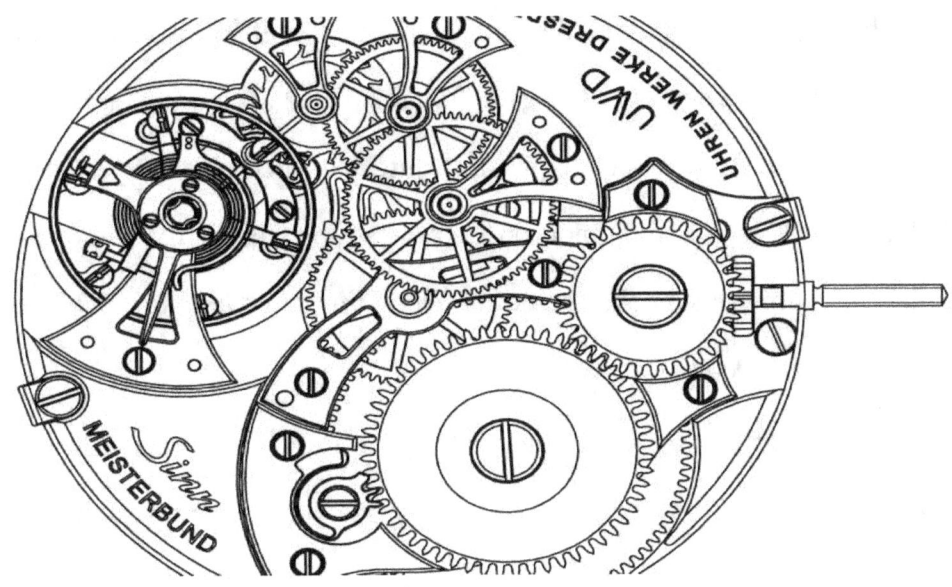

The watchmaker analogy is often used by the proponents of "intelligent design" to prove the existence of an active God at the helm of evolution. The premise is as simple as it is elegant. The intricate design of a watch evidences the existence of a watchmaker. A cruder but just as compelling analogy is to disassemble the components of pocket watch and deposit them willy nilly in a paper bag. Is it conceivable under any enormous stretch of imagination that by shaking the bag an infinite number of times that the parts would spontaneously assemble into the perfect timepiece from which it came?

The watchmaker analogy (1743 to 1805) was initially contrived by English Clergyman William Paley who maintained that a personal God was responsible for the design of things in nature.[xx] That a watch or human eye could assemble itself by chance is still touted by such Neo Darwinists as Richard Dawkins – a champion of the atheist and materialist camps and author of numerous books including *The Blind Watchmaker*, *The God Delusion* and *A Devil's Chaplain*.

Random mutation within the context of Natural Selection is Dawkin's analogue to the Blind Watchmaker.[xxi] Any sense that a Divine Watchmaker is up there designing intricate time pieces or human beings, is strictly fantasy, according to Dawkins and his "goose stepping" Neo-Darwinists. As Dawkin's framed the issue:

> "As an academic scientist, I am a passionate Darwinian, believing that natural selection is, if not the only driving force in evolution, certainly the only known force capable of producing the illusion of purpose which so strikes all who contemplate nature."[xxii]

Of course, the founder of the Blind Watchmaker feast is none other than Charles Darwin (1809 to 1882) and his seminal work – the *Origin of the Species* in 1852. Based on his extensive observations of birds on the remote island of the Galapagos Islands, Darwin advanced the theory that organisms evolved over huge stretches of time by a principle known as natural selection. Simply put, life was engaged in a process of intense competition marked by the survival of the fittest. Those biological organisms best adapted to the environment survived and were more likely to reproduce and populate the earth than the vanquished forms relegated to the fossil record. The first paleontological digs spurred by Darwin and his Theory of Evolution uncovered extensive evidence of evolution at work accentuating the development of the modern horse over time from more primitive forms to the evolution of man itself.

CHARLES DARWIN IN 1854

Darwin rejected any notion whatsoever that a preexisting God or Deity was involved in the evolution of life. As he framed this issue in a letter to Asa Gray in 1860:

> "I cannot persuade myself that a beneficent and omnipotent God would have designedly created the Ichneumodidae with the express intention of their feeding within the living bodies of caterpillars."[xxiii]

To this day, the proponents and antagonists of evolution occupy the outer reaches of the countervailing poles of evolution vs. intelligent design or some derivation of a deity involved in the origin of the species and mankind itself.

This author seeks to throw a monkey wrench (no evolutionist pun intended) into the machinations of both camps. To the evolutionists, the advent of genetics and molecular biology negate the core driving force of evolution as a credible force, to wit – natural selection. As will be detailed, it is mathematically impossible that life and complex forms could have evolved through a random process characterized by natural selection. On the flip side, it is equally disconcerting that a God supreme preordained the design of life on earth which underpins the concept of intelligent design.

Instead, it is far more probable that the progression of forms from the basic to intricate is emblematic of a universal principle of complexity as advocated by Teilhard. But unlike Teilhard who would ascribe to the intelligent design paradigm, the more plausible explanation is that such intelligent design is but a manifestation of an *"intelligent process"* propelled by a yet undiscovered "Law of Complexity" which propels structures inorganic and organic into more complex iterations.

In support of this suggested paradigm which this author coins as "intelligent process" is the complete failure of natural selection to account for the observed complexification in life.

4.1 NATURAL SELECTION & THE EMPEROR'S NEW CLOTHES

There's an old fairy tale by Hans Christian Anderson which tells the story of a powerful emperor who orders his chief tailor to create the most elaborate suit ever crafted. After receiving a King's Ransom to produce this finery and allegedly toiling for many weeks, the tailor presents the sovereign with his masterpiece wrapped in silk and adorned with ribbons of gold and silver. Upon uncovering this offering, the emperor is astonished to observe that the package is – *empty*. With a look of total disgust, the emperor immediately summons the royal guard to haul this "common thief" away, however, the tailor interjects – "Sire, don't let your eyes deceive you! The clothing is simply invisible to anyone who is treacherous, incompetent, or stupid." The emperor pondered that for a moment, closed his royal eyes, and once again gazed upon the tailor's present only to discover that he had – in fact - produced the most splendid attire ever crafted in the kingdom!

Thereafter, the emperor summoned the royal court and sought their agreement. Without a dissenting vote and unplanned excursion to the gallows – all agreed that the tailor had delivered on his promise to the King. And so, the Emperor paraded across the kingdom in a state of regal nakedness to the befuddled adoration of his subjects unwilling to appear incompetent, stupid, or otherwise unfit. The charade continues until a young child without any bias, pretense, or vested interest blurts out – "Why the Emperor has nothing on at all!" The facade having been exposed, the crowd concurs and the truth is finally revealed to the Emperor who then proceeds to render his so-called tailor as invisible as his creation.

The moral of this tale is that people should be willing to speak up and tell the truth even if they are afraid others will laugh at them. A more scientific moral is to avoid reaching a grand conclusion without the existence of empirical evidence.

I state without embellishment whatsoever, that the so-called theory of natural selection as touted by Charles Darwin is as devoid of truth and naked as the emperor's new clothes. A bevy of scholars, educators, and scientists of all ilk's have been unwilling to speak to truth in fear of being scorned and rejected by others. Since the pronouncement by Charles Darwin in 1859 that natural selection is the driving force of evolution, this vacuous notion has paraded about the most esteemed academies unchallenged and pristine. Quite the contrary, anyone who challenges this idea is considered uneducated, misinformed, or stupid. Sound familiar?

4.2 NATURAL SELECTION DOESN'T ADD UP

$$1+1=3$$

The thing about any theory is that it can be proven mathematically correct or not. When the element of scientific certainty is supplied – a theory is transformed into a law. For example, Isaac Newton's First Law of Motion is as follows:

> **"Every object in a state of uniform motion tends to remain in that state of motion unless an external force is applied to it."**

As an example, if I hit a tennis ball over a net there is a 100% chance that it will make it over to my opponent's side. There is a 0% chance that it will suddenly stop midair and levitate over the tennis court. If that is "Moster's Theory of Tennis Ball Levitation" – it will fail dismally when put to the test. I, of course, can broadcast with zeal my Tennis Ball Levitation Theory to the misinformed and even make a dollar or two with a good press agent and book release. I might even get invited to be a guest on "The View".

However, getting on a morning talk show will not elevate my theory to a law. I will never gain the prominence of Isaac Newton. And so, it is with the Theory of Natural Selection and it fares no better than my tennis ball.

The Theory of Natural Selection is 100% based on the notion that random variation can lead to large evolutionary changes. As an illustration, let's start with the evolution of the horse.

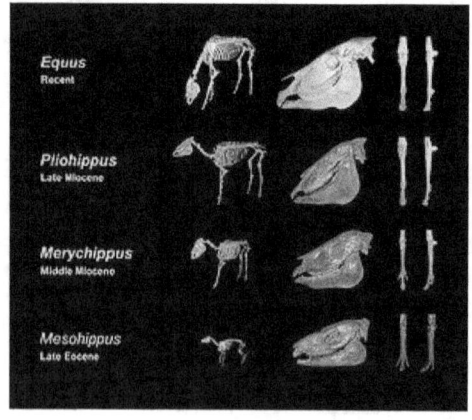

EVOLUTION OF THE HORSE

Analysis of the fossil record provides irrefutable evidence that the modern horse evolved from a progression of more primitive forms. To the dismay of the pure creationists, this author wholeheartedly endorses the phenomena of evolution. That said, the ultimate question is what is the driving force of evolution? Is it the aggregate culmination of nature throwing the dice – or is something else ahoof?

Dr. Lee Spetner examined this precise issue from a probability standpoint and natural selection came up dismally short *every time*. With regard to the probability of random variations leading to even one successful step in the evolutionary chain of our subject horse, Dr. Spetner concludes that the chances were *less than impossible*.

> "For just a moment let's look at the chance of a species evolving into a new one if at each step there is only one potential copying error that can be adaptive. What we've found above is the chance is just one of the small steps occurring. To get a new species, 500 of them have to occur without any failures. As we shall soon see, for successful evolution the probability of each has to be nearly one. The chance of 500 of these steps succeeding is 1/300,000 multiplied by itself 500 times. The odds against that happening are about 3.6 x 10 to the 2,738th power to one, or the chance of it happening is about 2.7 x 10 to the 2,739th power. That is a very small chance! It's more than 2000 orders of magnitude smaller than the chance of the event I called impossible." (emphasis in original)[xxiv]

Dr. Spetner also considers the total failure of scientists to prove that inert chemicals could ever evolve from the inorganic to organic.[xxv] This is a massive dilemma for Darwinists as the forces of natural selection would logically influence the development of inorganic material. Query, what would be the survival advantage of non-biological material to acquire the characteristics of a living organism? This huge stumbling block is sufficient to render the driving force of Darwinian evolution null and void.

Dr. Spetner is not alone in his critical analysis of Darwinian evolution. British-Australian biochemist, Michael Denton, has also written extensively on this topic most recently critiquing his seminal work, Evolution: A Theory in Crisis.[xxvi]

"There has been massive advances and discoveries in many areas of biology since Evolution was first published. These developments have transformed biology and evolutionary thought. Yet, orthodox evolutionary theory is unable to explain the origins of various taxa defining innovations. This was my position in *Evolution*. It remains my position today."[xxvii]

Denton considers several examples of extreme morphology wherein changes in biological structure cannot be attributed to natural selection. The wings of a bat are a prime example.

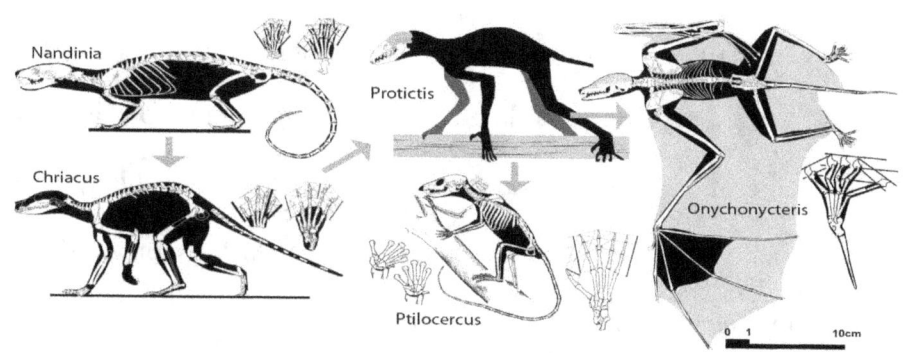

BAT EVOLUTIONARY TREE

He observes that the first winged bats appeared in the fossil record 50 Million Years ago. According to Denton, adaptive forces outlined by Darwin cannot explain the morphological changes bridging the pre-flight bat limb and wings which ultimately emerged. He cites to research and conclusions reached by biologist Glen Jespen for the following proposition:

"No one has successfully proposed a process which would be effective in the change from one niche to the other; whether the bridging group would be pulled by advantages of the new milieu or pushed by disadvantages of the old."[xxviii]

Denton delves into the enormous genetic and molecular changes which would be needed to transform the bat limb to wing.

"First, we found 14 genes that are likely associated with digital elongation in bats – two bx genes (Tbx3 and Tbx15), five genes from

BMP pathway (Bmp3, Rgmb, Smad1, Smad4, and Nog), for Homebox genes (Hixd8, Hoxd9, Satb1, and Hota1) and three other genes (Twist1, Tmeff2, and Enpp2) related to either digital malformation or cell proliferation. Next we identified seven genes that are likely associated with morphological and functional similarities between the thumb and hind limbs digits."[xxix]

Denton concludes that natural selection in light of the facts cannot be justified. "Morphological change requires extensive genetic rewriting."[xxx]

Therefore, the probability of even one step occurring in the evolutionary progression of a bat or horse is virtually identical to that of my tennis ball levitating over the tennis court. It is less than impossible. This same zero likelihood results whether we are considering the evolution of a bat, horse, or a human being.

All things being less than zero, therefore, shall be draw the conclusion that evolution is a nullity? Absolutely not! The paleontological fossil record has demonstrated that the evolutionary dots can be connected from the primitive to the modern iteration of whatever species where are observing. I would maintain that evolution as a process has transcended from a mere theory to a law. The HUGE caveat, however, is that randomness and its stepchild "natural selection" are not driving it. There is another force at work which, as described above, manifests in everything from digital e-life creations to paleontology to physics and cosmology. Teilhard coined the term "*complexification*" and he is correct in my estimation. The "Theory of Complexification", therefore, is a far better explanation of the driving force of evolution and analogous phenomena than the "Theory of Natural Selection". But does such conclusion contradict any possible notion of a divine energy or God force as theorized by Teilhard. Certainly not.

4.3   CREATIONISM ALSO DOESN'T ADD UP

Creationism dictates that an omnipotent and omnipresent God created the universe along with its bounty of life in all of its diversity and splendor. According to the biblical account, the Earth was created a mere 6,000 years ago and humankind emerged in a garden populated by mostly divine artifacts with the exception of one really malevolent serpent. With just a modicum of scrutiny, the creationist theory fares no better than natural selection.

It has been proven to a certainty that the actual age of the earth is 4.543 Billon years – just a tad bit greater than 6,000. In a process known as radiometric dating, scientists measure the geological age of radioactive deposits such as uranium by observing the process of decay of radiation over an extended period of time. The depletion of a sample of uranium by a factor of 50% is referred to as "half-life". In the case of uranium – the half-life has been determined to be 4.5 billion years – which is the measure by which the age of the earth is established. Over 40 variations of this test have been performed and verified by scientists worldwide and the results are in. Alas, the 6,000-year age predicted in the bible is dead wrong.

Along with this revelation, so to speak, we can also discredit the pseudo-scientific reports that evolution is a fake and that dinosaurs never existed or alternatively, happily shared the Jurassic Park with humankind.

All of these so-called findings have been discredited. Simply put, the likelihood of early man and dinosaurs coexisting on this planet – is zero. It is as incredulous as – well – natural selection. It didn't happen.

Creationism as a dogma has also had a devastating effect on the freedom of scientific thought and the personal longevity of many advocates with a contrary view. To find the stultifying and terrifying evidence of same, look no further than one of the greatest minds of all time – Galileo. His observations using a new device known as the telescope led him to believe that the Earth was not the center of the universe as espoused by the Church but revolved around the Sun. This conclusion conflicted with the biblical account and was not appreciated by the Vatican, to say the least. On April 12, 1633, Father Vincenzo Maculano da Firenzuola, appointed by Pope Urban VIII began inquisition proceedings against the astronomer and ordered him to report to the Holy Office. On June 22, 1633, the Church handed down the following order:

> "We pronounce, judge, and declare, that you, the said Galileo… have rendered yourself vehemently suspected by this Holy Office of heresy, that is, of having believed and held the doctrine (which is false and contrary to the Holy and Divine Scriptures) that the sun is the center of the world, and that it does not move from east to west, and that the earth does move, and is not the center of the world."[xxxi]

TRIAL OF GALILEO

Seeking to avoid the physical separation and rotation of his head from the literal body of the Church, Galileo recanted, although it is rumored that his last words before departing for all thing's celestial was the irksome remark – "and yet it moves!"[xxxii]. It took the Church over 300 years to admit that Galileo was right along and that scientific inquiry had not been given a fair shake.

The danger of the fallacy of Creationism is that it constructs what appears to be an impenetrable barrier and separation between the concept of a divine God force as manifesting in the universe and man – and the laws of science. Such leads to the equally fallacious views of the pure Creationists who deny the existence of evolution and the equally malignant atheists who espouse precisely the opposite as epitomized by Richard Dawkins. It is the fervent view of this writer that both extremes are incorrect.

As further discussed, Teilhard himself fell victim to the human centrist perspective of Creationist thinking, notwithstanding his advocacy of scientific thought and evolution itself. His lifelong commitment as a Jesuit led Teilhard to embrace a human paleontological centrism reminiscent of those church officials who condemned Galileo for heresy. Evolution revolved around mankind. Our species was the center and pinnacle of the evolutionary progression. In less scientific parlance, we were the cat's meow.

John A. Roebling (1806 to 1869) devised an engineering solution which his contemporaries deemed impossible – how to span the banks of the East River in New York with a bridge connecting Manhattan and Brooklyn. The Brooklyn Bridge which was officially opened to the public on May 24, 1883 was dubbed the "Eighth Wonder of the World". And it still is. This feat of imagination and engineering is the perfect metaphor for the task now before us.

There appears to be an implacable gap between the core concepts of evolution and intelligent design (Creationism). Teilhard's study of the *Phenomenon of Man* provides the most useful paradigm for constructing a workable bridge between these two extremes. Of the myriad of theories advanced by Teilhard, it is the principle of *complexity* which offers the great opportunity to systemically connect the scientific and theological. Indeed, an analysis of complexity within the realm of evolution, cosmology, and physics creates the pillars for the greatest extension bridge of the human mind. Let us now embark on the construction project which will rival the engineering feat of John Roebling himself! Mr. Roebling's solution to the success of the world's longest suspension bridge was the use of metal cables. Our connective construct will serve the same function between the banks of the scientific and theological. Complexity will be our steel cable.

OPENING OF BROOKLYN BRIDGE – MAY 24, 1883

CHAPTER FIVE

COMPLEXITY AND THE EVOLUTION OF MAN

5.1 COMPLEXITY AND THE HUMAN BRAIN

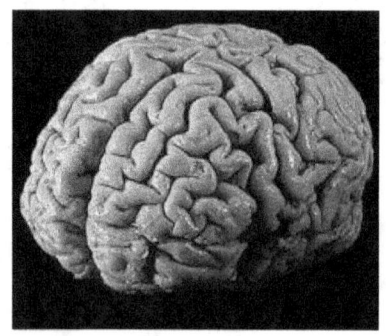

The evidence is irrefutable that the size and complexity of the human brain has increased over time. The fossil record demonstrates such a progression spanning millions of years of development. During the course of approximately seven million years, the human brain has nearly tripled in size. Although branches of the primate group appear to have reached a stable and consistent level of cranial size – for example – apes and chimpanzees – the human brain has continued its upward climb. The paleontological evidence demonstrates that the evolution of man is marked by a measurable increase in the size and complexification of the brain. You just need to look at the fossil evidence which is indisputable.

One of the earliest pre-human fossils discovered, *Australopithecus afarensis*, a/k/a "Lucy had skulls with volumes of between 400 and 550 milliliters which is actually within the cranial range of chimpanzees but less than gorilla (700 ml)."

AUSTRALOPITHECUS AFARENSUS – "LUCY"

Lucy lived 3.2 million years ago and was discovered by Donald Johanson in Eastern Africa. Johanson celebrated this astounding discovery by singing several choruses of the Beatles "Lucy in the Sky with Diamonds". Somehow the name "Lucy" stuck and christened this pivotal discovery. Little is known about the behavior of Lucy (or Ricky for that matter) and there is no evidence as to the development of tools or any artifacts indicative of higher intelligence. Skeletal studies and fossilized footprints strongly indicate that Lucy stood upright. However, she sported apelike proportions, a flat and protruding face, and curved fingers well adapted for climbing trees.

The next rung in the evolutionary development was a more modern hominid – Homo Habilis – which appeared about 1.9 million years ago. The brain size topped off at 600 ml which was a substantial increase in capacity compared to Lucy. Scientists also believe that there was a corresponding increase in the size of the frontal lobe of the brain linked to the development of language.

HOMO HABILIS

Paleontologist legends, Louis and Mary Leakey discovered the fossils of this pre-human between 1960 and 1963 in Tanzania along with thousands of crude stone tools. This evidence of what was believed to be the first use of tools was highly significant because it correlated with the substantial increase in the size of the Homo Habilis brain which was 50% larger than its predecessor Australopithecus.

The upward trajectory of Teilhard's arrow of complexity continued with the appearance of the new pre-human species Homo Erectus approximately 1.8 million years ago to 70,000 years ago. Once again, an increase in brain size corresponded with new technological developments, i.e., the first reported use of fire and further refinement of stone tools and spears.

<u>HOMO ERECTUS</u>

With the arrival of our own species, Homo Sapiens, approximately 500,000 years ago, the size of the brain increased substantially to over 1300 ml or more than twice the capacity of Homo Erectus. Neurologists have pinpointed not only the increase in cranial size but complexification of critical neural regions such as the frontal lobe responsible for the development of language and the formulation of abstract concepts. This history of human development and technology as compared to the crude but significant achievements of Homo Erectus has been nothing short of supersonic. Our technology and scientific achievements have literally transformed the planet and propelled our artifacts and imagination to the moon and now interstellar space itself.[xxxiii]

HOMO SAPIEN ON LOCATION

The evidence of complexification of the human brain from its primitive origins 3.2 million years ago to present is irrefutable. This documented scientific upward progression in the size and complexity of brain development is a fact and not the product of conjecture or idle speculation. Critical to the thesis of this essay, the evolution of the human brain has traced the upward trajectory of an arrow always pointing in the direction of greater complexity. As will be seen later on, this conclusion does not fly in the face of the existence of a higher power or God – it lays the very foundation for its emergence and the construction of a formidable bridge connecting the scientific and theological. John A. Roebling would be proud indeed.[xxxiv]

5.2 COMPLEXITY AND THE EVOLUTION OF PLANTS

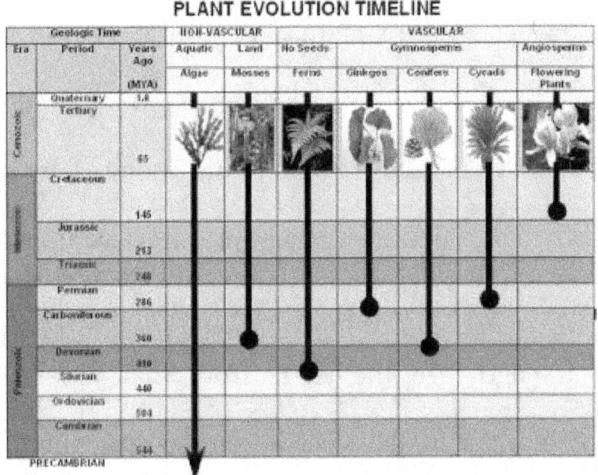

PLANT EVOLUTION TIMELINE

For further evidence of complexification and its decided upward progression look no further than your backyard or neighborhood garden. The evolution of plants is once again irrefutable evidence of evolution and the driving force of complexity over the eons. The earliest appearance of plant life in the form of microscopic algae appears in the fossil record more than 3 billion years ago. The process of photosynthesis central to the development of plants made its first appearance in the early oceans approximately 2.5 billion years ago. During the next billion years, plant life increased in diversity but remained at the unicellular level. Shortly thereafter, however, the first multicellular plants emerged. The next massive milestone occurred 450 million years ago when the first land plants arrived on the scene. Approximately 60 million years later during the Devonian period, the first ferns and seeds emerged followed by the familiar structures which still characterize plant life today – leaves, roots, and branches. And what Valentine's Day would be complete without the presentation of a bouquet of roses – flowering plants, which first appeared 145 million years ago. In the years from that time to present, the complexification of plant life has created the natural palette celebrated by such masters as Monet.

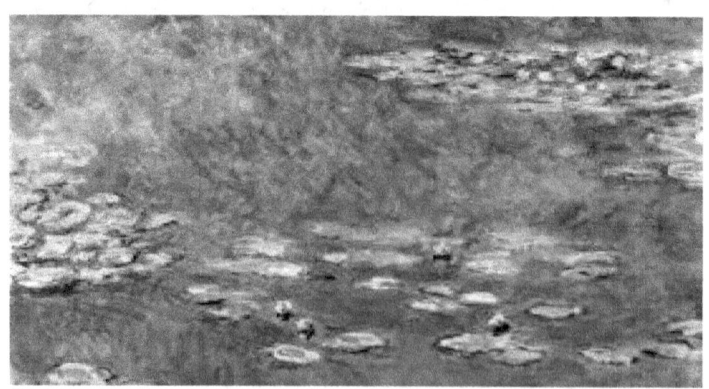

CLAUDE MONET'S WATER LILLIES

The evolution of plant life from its unicellular origins to primeval forests is a fact. Once again, the upward progression of plant evolution traces the immutable direction of an arrow pointing toward greater complexity over time.

5.3 COMPLEXITY AND THE EVOLUTION OF THE HUMAN EYE

Having viewed complexity and evolution from the macro level of human brain and the development of complex plant life, let us now peer inward to examine the human eye, the evolution of which – has enabled us to peer in the first place.

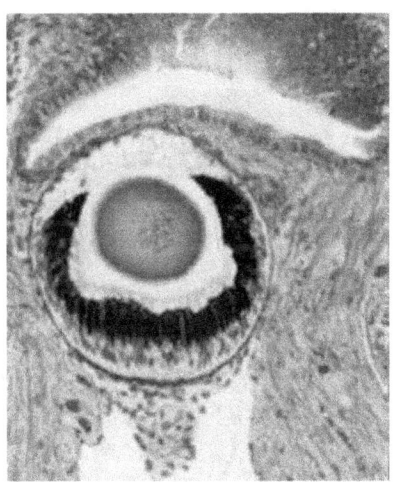

PRIMITIVE LENS FOUND IN GARDEN SNAIL

The evolution of the human eye in all its complexity troubled Darwin himself. Although Darwin's reservations have fueled Creationists in their attempt to negate the existence of evolution, it was not the process of evolution which troubled Darwin, only its mechanism, to wit, natural selection.[xxxv] The author whole heartedly agrees that natural selction which is driven by random mutation could not have explained the increase in complexification of the eye over the eons. However, as set forth herein, no one can question that complexification of eye structure from its primitive origin to the human eye *did occur*. Natural selection as devised by Darwin cannot account for complexification whether one considers the development of the human brain, a flowering plant, or your own eyes. However, that does not negate the existence of evolution as a fact and that some process resulting in complexification is actually occurring, although its scientific underpinning has yet to be crystallized in the early 21st Century.

The earliest predecessor of eyes emerged approximately 700 million years ago in the form of unicellular organisms known as "eye spots" which could sense the presence of light or absence thereof.[xxxvi]

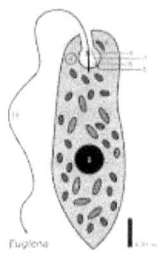

<u>EYE SPOT IN SIMPLE ORGANISM</u>

During the course of the next 200 million years, the eye "spots" morphed into receded areas or "pits" which allowed for greater receptivity to light and its refracted images. The first fossils of eyes with similarities to our own emerged during a period of rapid evolutionary development known as the Cambrian Period or Cambrian Explosion – 540 million years ago. During the Cambrian period, the complexification of the eye structure proceeded at a rapid pace. The fossil evidence is replete with the successive development of structure and complexity. The eye pit deepened in the form of "sockets" to allow for even greater light sensitivity, the lens of the eye emerged followed by the complex ocular structure present in mammals, primates, and ourselves.

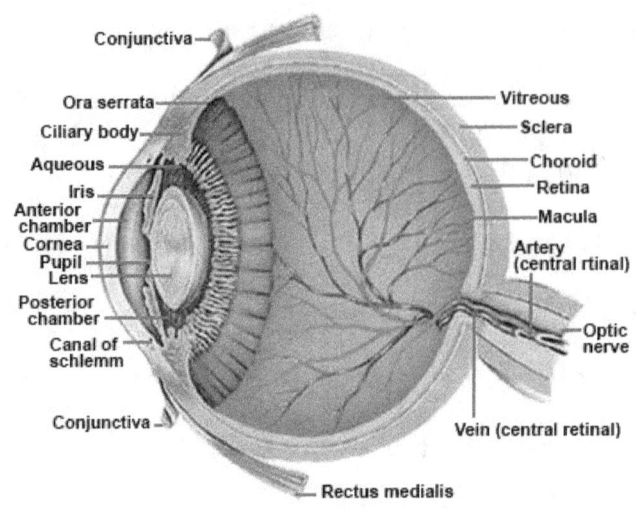

ANATOMY OF HUMAN EYE

The evolution of the human eye is direct evidence of evolution. However, it is the complexification of its development from the primitive appearance of eye spots to the variegated and intricate structure of the human eye which is the subject of this work. Whether we peer outside ourselves into the natural world or gaze upon our own eyes (and ourselves) in the mirror – the irrefutable evidence is that complexification is a real and verifiable phenomenon – an arrow which points ever upward.

CHAPTER SIX

COMPLEXITY AND THE COSMOLOGICAL RECORD

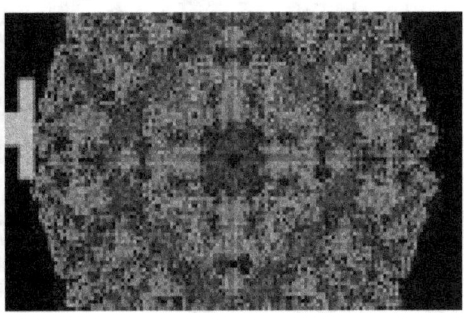

EVOLUTION OF UNIVERSE – COSMOLOGICAL & ELECTRONIC

You can easily mistake the graphic of cosmic evolution (left) with the computer-generated e-life patterns (right) discussed at length in Chapter 2. The trail of Teilhard's arrow is tracing its upward path to complexity in both realms.

Precise and verifiable astronomical research has documented that our universe arose in a massive explosion of pure energy approximately 13.799 billion years ago, give or take a few million or so. This theory originated by Edwin Hubble in the 1920's was based on his observations that distant stars and galaxies were moving away from earth in every direction and at an increasing rate of speed. As with many new theories, it took the scientific community and society at large many years to accept Hubble's explanation as plausible. Interestingly, its popular name today – the Big Bang – was coined by his critics as a way to lampoon the great astronomer. Over time, however, Hubble's theory of cosmic evolution and its name gained traction and almost universal acceptance.

Although there are comparisons with the advent of the Big Bang and Genesis, the bang was a bust from an explosion standpoint as there was no light or sound in the "beginning". The first phase of the Big Bang is also referred to the Planck Epoch and assigned a time factor of 10 to the -43st of a second after the Big Bang. During this period the first energy was so intense and heated that it cannot be replicated even in the largest particle accelerator in the world – the Large Hadron Collider at CERN Switzerland. Given the enormous temperatures, it is theorized that the four fundamental forces in the universe – electromagnetism, gravitation, weak, and strong interaction were combined as a unified force.

During the next period between 10 to the -43rd and 10 to the -32nd, the initial burst of energy cooled very slightly and inflated in size to a massive scale from a single point to distances in the millions of light years. Critically, the unified state of combined forces began the process of separating into the four critical forces noted above as the temperatures dropped. However, the temperatures were still off the measurable scales and atomic particles were unable to form.

Between 3 minutes to 20 minutes after the Big Bang, temperatures finally dropped to the level were the basic atomic particles such as protons and neutrons were able to arise and form atomic nuclei.

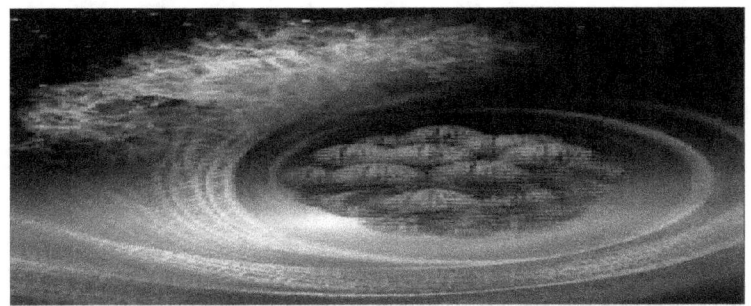

ILLUSTRATION OF ATOMIC NUCLEUS

Even a cursory view of the cosmological evidence demonstrates the ascending arrow of complexity from a single point – to pure energy – to centralized structure.

During the so-called "Dark Ages" of cosmic evolution, the universe continued to expand and cool, but had not dropped to the point where light in the form of photons could be released. Therefore, this early universe was literally invisible then and still is today as no light was released which could be captured by human telescopes billions of years after the Big Bang.[xxxvii]

The Cosmological Dark Ages came to an end approximately 400,000 years after the Big Bang when photons were finally released and there was a massive release of light which would do Hollywood proud along with the proponents of a literal interpretation of Genesis. Certainly, very near the Beginning of all things, there was light.

With the advent of light and cooling temperatures, the first structures in the universe emerged including stars and quasars.

**IMAGE OF WHAT MAY BE FIRST GALAXY
(490 Million Years after Big Bang)**

Data collected from the Hubble Space Telescope establishes that the initial small number of galaxies combined into larger galaxies when the universe was only 5% of its current age. Our own Milky Way Galaxy formed 8.8 Billion years ago followed by the birth of our Solar System.

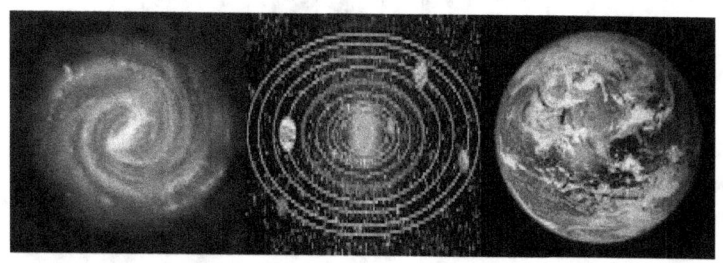

MILKY WAY GALAXY, SOLAR SYSTEM, AND HOME SWEET HOME

Our Solar System coalesced approximately 4.6 Billion years ago from a contracting molecular cloud which first gave birth to the sun in an explosion caused by the forced atomic combination of hydrogen into helium in a process known as fusion followed thereafter by our own planet earth and its siblings.

The evidence is indisputable that our universe evolved from the most basic components to the building blocks of matter – atoms – to early stars, galaxies, solar systems, and finally planets. Superimpose this progression with the successive development of electronic life, the fossil record, and evolution of the human eye and only one conclusion is possible – a process of complexification has occurred in all things. Teilhard's Arrow is for real.

CHAPTER SEVEN

COMPLEXITY AND THE ATOMIC RECORD

THE LARGE HADRON COLLIDER

Our modern equivalent of the great Pyramids is buried deep beneath the ground in Geneva, Switzerland on the French border. It is the largest and most complex machine ever built by man and has a circumference of over 17 miles. The Large Hadron Collider also known as the LHC, is comprised of a ring of superconducting magnets which boost two high energy particles in a race around this atomic circle at velocities close to the speed of light. In the atomic equivalent of a car crash, the two beams of high energy particles collide with each other at phenomenal speeds. The aftermath of this atomic crash is analyzed by atomic scientists who – no doubt – loved "Hot Wheels" toy racing cars as kids. The broken and scattered pieces of atomic particles – subatomic particles and even more minute components – are analyzed by the world's most powerful computers to provide a glimpse into the unseen world of subatomic structure.

What the LHC has uncovered is miraculous. From the infinitesimal stirrings of atomic structures which seemingly appear and disappear in a trillionth of a second, to more stable particles, to atoms, to molecules, to elements, to visible matter, and then us – the atomic world is nested in levels of increasing complexity. Sound familiar?

The LHC has provided direct evidence along with its less powerful predecessors of the existence of what is referred to as the Standard Model of Physics. To date, physicists have discovered 61 elementary particles including 36 quarks, 12 leptons, 8 gluons, and a myriad of other exotic cousins.

On July 4, 2012, physicists at LHC celebrated their independence by announcing the discovery of the elusive "Higgs Particle" – sometimes referred to as the "God Particle" which was theorized to existence since the early 1960's. The Higgs Particle is also referred to as the "Higgs Field" and is responsible for giving particles their mass, absent which – the universe would never have emerged beyond a state of pure energy.

LHC COMPUTER DISPLAY REVEALING EXISTENCE OF HIGGS PARTICLE

The significance of the detection of the Higgs Particle and the plethora of particles which comprise the Standard Model is that what we perceive as matter – for example – the physical condition of everyone reading this sentence – is a hierarchical structure of Russian Dolls – each one nested within the other and ascending a scale from the infinitesimal to the massive. Once again, we have confirmation that the arrow of complexity exists and it permeates the atomic world from its essence to us.

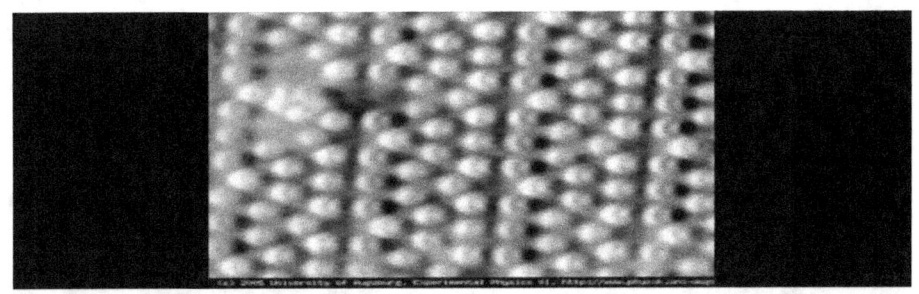

PHOTOGRAPH OF ACTUAL ATOMS AT THE MOST ELEMENTARY LEVEL

The Jury Verdict is out and Complexity as a verifiable force of nature exists whether we are observing the LHC evidence of the Higgs Particle or the emergence of the first galaxies at the edge of the universe. Critically, evolution as it is manifested across the wide spectrum of disciplines from paleontology to physics to biology is but the progression of an unbroken and nested chain of Russian Dolls rising in greater and greater levels of complexity and driven by an underlying principle which defies Darwin's simplistic notion of natural selection. Evolution has a direction and as intuited and theorized by Teilhard, it is ascending towards the heavens – an arrow pointing and beckoning all of us in the direction of the unseen and unknowable which is best described by the word God.

CHAPTER EIGHT

TEILHARD'S ARROW – YOU ARE HERE

We all want to desperately believe that we are unique and self-important. When it comes down to the very foundation of human existence, we emerge from the birth canal alone and will exit the vehicle of human existence in the identical solitary state. Given this quandary, it is not surprising that almost all of us have a myopic view of the universe which typically puts us at the center. Galileo, to the dismay and condemnation of the Church, scientifically proved that such ego-centrism was flawed and that Mother Earth and its children simply revolved around the sun. In fact, the sun and every other celestial body including our own Milky Way Galaxy revolves around something else in a nested progression of the ubiquitous Russian Dolls.

The operative effect of the obvious conclusion that we are not the center and rationale for the existence of the universe is unnerving to say the least. Such determination has resulted in the great contortions of logic and belief on both sides of the spectrum whether purportedly scientific or theological. On one side of this extreme, the atheists proclaim that life emerged from random interactions and that there is no direction or purpose in our individual lives. The Neo-Darwinists such as Richard Dawkins have famously and articulately illuminated this argument. The flip side, of course, are the Creationists who view the bible as the literal word of God and espouse that the first chapter thereof provides all the explanations we will ever need.

My personal hero, Pierre Teilhard Chardin, unfortunately, seems to be straddling an untenable position somewhere near the middle of both camps. Although he brilliantly sets forth his argument for the existence of evolution and its direction toward greater complexity, the human-centric bias persists as mankind is placed literally at the pinnacle of evolution. Indeed, all of evolution and its complexity was intended to result in the culmination of the most ingenious life form every conceived – us. To Teilhard, there is definitely an arrow in the progression of evolution and mankind's precise GPS coordinates are on its very edge.

Although it is psychologically beneficial to have a palpable sense of self, science instructs that we cannot be the cause celebre of all things. That admission, however, does not negate the existence of a higher power or God, but affirms that the divine presence **must** co-exist with mankind. Teilhard's ultimate centric vision of God Jesus – Omega – could not have preexisted mankind, but emerged as part of an evolutionary process which is still ongoing. Omega is an unseen and unknowable point beyond where mankind is situated on Teilhard's Arrow. It co-exists with us at this very moment – but at a qualitative level beyond our ability to perceive it. And as the bible instructs that Jesus is the Son of the Father, such is validation that even at this sentient and miraculous level of consciousness and creation, there is a point of complexity even beyond "Him".

8.1 THE COPERNICAN PRINCIPLE

NICOLOUS COPERNICUS

Nicolaus Copernicus (1473 to 1543) courageously stated that the earth is not the center of the solar system or universe. Although Galileo is credited with astronomical observation to prove Copernicus correct, good ole Nicolaus should be granted a post-humus Noble Prize for getting it right in the first place.

The Copernican Principle, named after the founder himself, asserts that there is nothing unique about our individual human perspective and that we are statistically unlikely to observe anything at either the beginning or end of its existence at a given moment - whether we are observing a Broadway Play or the probabilistic likelihood of dying at the very moment this sentence was written or when you the reader are digesting same. The rule states that there is a 95% chance that an observer is NOT at the first 2.5% or last 2.5% of anything which is being examined.

Therefore, when applying the Copernican Principle to mankind's own location on Teilhard's Arrow of evolution and complexity, there is a 95% chance that you and I are not sitting on the edge of the arrow and alas – NOT the pinnacle of evolution or creation – whatever your perspective might be.

Such is also a rational conclusion which could also be derived from the observation of Teilhard's arrow at any point in the continuum whether paleontological, cosmic, electronic, biological, or atomic. Ask yourselves this most logical series of questions:

1. What was the likelihood that any of our pre-human predecessors whether homo habilis or homo erectus were situated at the apex of evolution?

2. What is the likelihood that the latest electronic manifestation on your computer screen, say the R Penetomino is the *final* iteration of electronic life? Remember that curious e-creature which evolved at generation 33,699,586 (Gemini)?

3. What is the likelihood that any cosmic structure in the universe is the final stage in the evolutionary process – for example, our own planet earth? Astronomical observations have demonstrated that the stuff of the universe is in a constant state of flux and change from one state to the next.

4. What is the likelihood that any particle discovered by physicists in the ingenious LHC or some miraculous device with a million times its power to follow – is the very last particle in the progression of matter from the Higgs to the Quark to the Electron and so on?

The logical answer to the above questions is obviously "zero". Therefore, whether we are viewing the location of mankind on Teilhard's Arrow within the statistical confines of the Copernican Principle or simply straight logic, there is almost a zero chance that mankind is sitting right now on the very edge of evolution.

That said, it is more likely than not, that mankind can be placed at some point other than the first or last 2.5% of Teilhard's Arrow, most likely near the center. That being the case, it is more likely than not, that at this very moment, there exists some greater iteration of complexity than you and I at a point beyond our current placement along Teilhard's Arrow.

Further validation of this principle is the fact that nature does not extinguish every predecessor form of life in the evolutionary continuum. Species which existed millions of years ago continue on unabated. Examples include certain species of sharks (unchanged for 200 million years), lampreys (360 million years), Coelacanth[xxxviii] (360 million years), Horseshoe crabs (445 million years), nautilus (500 million years), and sponges (760 million). Therefore, biological organisms of lesser complexity continue to coexist with their descendants of much greater complexity. For example, jawless fish (like the lampreys or coelacanths) predated the evolutionary development of jawed fish from which all vertebrates with "faces" emerged including reptiles, mammals, and us.

Lampreys, although occupying an earlier position on Teilhard's Arrow, coexist circa 2017 with swordfish, salmon, and other species.

COEXISTENCE OF ANCIENT AND MODERN FORMS COELACANTH TO SWORDFISH TO REPTILE

The onrush of the evolutionary wave does not always replace all prior structures which continue on in the wake of emergent and modern structures. In the same vein, there is no logical reason to exclude this phenomenon of the coexistence of ancient and emergent forms from characterizing our own relative position as Homo Sapiens on Teilhard's Arrow. Is it not possible that we are the primate analogue of a jawless fish long since superseded by a more complex descendant further along the continuum of Teilhard's Arrow? Point of fact is that the evolution of organic forms is *always a moving target*. One life form or another has always superseded another. That being the case, why would mankind occupy the unique position of being at the very edge of the breaking evolutionary wave crashing into the future? Clearly, there is no logical basis for such deduction. It is thus more likely than not that Homo Sapiens coexist with more complex successors at a further point along Teilhard's Arrow. Right now.

Seeking further validation (at the risk of redundancy), one need only look at the progression of those Russian Dolls again. Is it not the underlying structure of cosmology, biology, and physics that the subject of our observations is but nested Russian Dolls moving forward in complexity? It is, therefore, more likely than not, that you and I (at our existing point on the continuum of Teilhard's arrow) are but a component part of a more complex structure of which we are a lower level nested part.

The grand conclusion of this analysis, therefore, is that whether we apply a statistical measure or simply deduce our relative position on Teilhard's arrow based on the objective evidence of the paleontological fossil record or most recent observations of the LHC – there are nested structures of less complexity and greater complexity which co-exist with us at this very moment.

CHAPTER NINE

THE FACE OF GOD

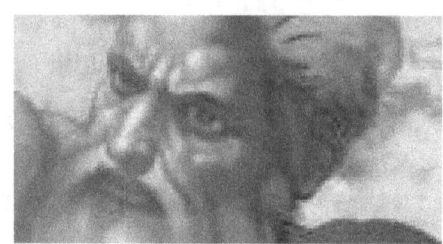

The Trillion Dollar Question, therefore, is whether a Higher Power or God really exists or whether it is a convenient crutch or rationalization to fend off the fear of death or existential uselessness? It is my ardent belief that scientific process and logic lead to the inescapable conclusion that God does exist and that at this very moment all of us co-exist with God. We are the jawless fish equivalent to our own sentient successor which exists contemporaneously with us. This conclusion must be borne out from our scientific observation and logical conclusions.

That said, the truly fascinating question is – "What are the characteristics of this Higher Power and God? Can we actually see the "Face of God" or scientifically extrapolate the divine features of the Lord? I believe the latter is the correct assessment for the following reason:

Successive points in the evolutionary progression contain emergent characteristics which cannot be predicted from a point of observation at a less complex point along Teilhard's Arrow.

As an example, from the world of atomic particles, let us consider three simple atoms – two of hydrogen and one of oxygen. Although we cannot see or perceive oxygen and hydrogen, we can scientifically determine its presence in the process of respiration or sailing above the Albuquerque skyline in a lighter than air balloon. However, if we take two hydrogen atoms of this invisible gas and combine it with one oxygen atom – the result is H2O or water. The physical properties of water are emergent features from the combination of the constituent elements and qualitatively distinct from the predecessor elements. By its very nature the term "Emergent" means that the resulting product is not the simple sum of its parts, but a product which is greater than the sum of its parts.

Assuming you and I were reduced to the atomic scale of an oxygen atom in a world that consisted entirely of subatomic particles, we would not be able to predict, yet alone observe, the presence of water which – in actuality – existed (or co-existed) with us at the identical time and place. It would occupy a realm or dimension of greater complexity beyond our more primitive vantage point.

Accordingly, it is more likely than not, that a consciousness at a point of greater complexity than mankind exists at this very moment at a further point along the evolutionary continuum or Teilhard's Arrow. As in the case of our hypothetical observers at the atomic level, we would not be able to observe or scientifically prove the existence of the emergent structure beyond the level of you and I as its properties would be greater than the sum of its parts. However, I do believe that Teilhard is right on the money with his hypothesis that there is a certification or coalescence of *thought* beyond the limited mentality of our individual brains. Teilhard considers this higher level – the noosphere - to be a greater consciousness which emerges from the collective components of all humankind. And keeping aligned with Teilhard's theory, there is likely an even greater level of complexity which exists here and now above and beyond the noosphere – which Teilhard terms as Omega or Christ Jesus. That such divinity exists is more likely than not from a strictly scientific perspective.

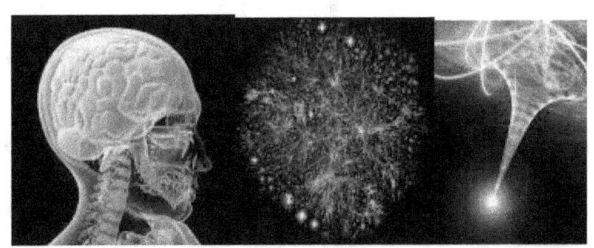

HUMANS TO NOOSPHERE TO OMEGA POINT

Therefore, it is more likely than not, that we cannot predict what the face of God would look like from our perspective as it would be an emergent construct further along in complexity. That said, I believe we could intuit or indirectly verify its presence in our life as humankind is a constituent element at a lower level complexity. However, our existence at such inferior level is essential to the emergence of God and his/her/its face – so to speak. We are lower nested structures which fit together like our ubiquitous Russian Dolls in the larger scheme of all things. And thus, this bring me back to Teilhard's quote which was cited at the beginning of this work:

> "If Omega were only a remote and ideal focus destined to emerge at the end of time from the convergence of terrestrial consciousness, nothing could make it known to

us in anticipation of this convergence. At the present time, no other energy of a personal nature could be detected on earth save that represented by the sum of human persons. If, on the other hand, Omega is, as we have admitted, *already in existence* and operative at the very core of the thinking mass, then it would seem inevitable that its existence should be manifested to use here and now through some traces." (Emphasis in Original)[xxxix]

Omega- Christ Jesus – Yahweh – the Face of God must logically and scientifically co-exist with us here and now. And what would the manifestations or traces be of such higher consciousness? Well – any person of faith could answer that without detailed reflection. The spirit of God, although we cannot see His face, manifests in everything we do. Distilled to its essence, we can sense and feel the spirit of God in all things. I believe that is what Teilhard refers to as "traces" of God. And given our systemic inability to perceive subsequent emergent characteristics from our more primitive perspective, we may never be able to see the Face of God and have to content ourselves with being receptive to the traces thereof.

CHAPTER TEN

INTELLIGENT PROCESS

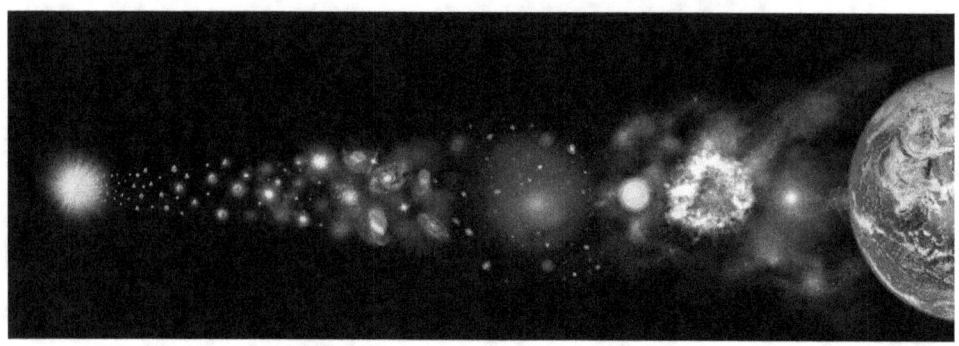

The term "Intelligent Process", as previously discussed, is an attempt by this author to bridge the gap between Teilhard's concept of evolution driven by a preexisting designer and the blind watchmaker notion which underpins Darwinian evolution. As previously discussed, both creationism a/k/a Intelligent Design and Darwinian Evolution suffer from identical infirmities. Creationism as a theory cannot advance to the status of scientific law.

Similarly, Natural Selection is – well – dead in the water and cannot *survive* scientific scrutiny. Complexity as evidenced by its manifestations in evolution, physics, e-life, and astronomy cannot be propelled by random interactions and the sophomoric pronouncement that the battle for survival is the ultimate answer.

Teilhard's hybrid construct is an attempt to move closer to the center between the divine and provable, but still requires the *deus ex machina* to descend from the heavens and provide the answer to the ultimate riddle of creation.

The solution as advanced in this essay logically fits together like those... well – Russian Dolls. The human brain and plants and biological structures and standard model of physics and all of us – were likely not the result of an intelligent designer who preexisted all of us. It is more probable than not based on the aforesaid evidence embedded in the paleontological, cosmological, biological, and atomic record, that our watchmaker and designer evolved along the same progression of evolution – Teilhard's Arrow – as we did and everything else on earth and above. A higher power or God or Christ Jesus or Omega resulted not from an intelligent designer but intelligent process, to wit, a higher level of complexity which emerged at a point beyond our less developed and finite location along the evolutionary continuum and Teilhard's Arrow. As this greater consciousness would co-exist with us in the here and now, it could easily lead to the erroneous view that it preexisted us – that in the Beginning there existed a greater consciousness – a power – a quantum potential which many refer to as God.

I absolutely believe in the existence of God, but as a consciousness that emerged from an intelligent process. And that deduction makes all the difference in the world to the central thesis of this essay.

Which brings us back to John Roebling and his amazing Brooklyn bridge. The greatest minds and industrialists of the time said it could not be built. That cables could not traverse the span between the two far off shores. However, Roebling persisted – he refused to give up and his brilliant bridge became a reality.

In the same vein, the great and distinct shores of the scientific and theological realms can also be spanned and connected. Evolution confirms the existence of a bridge and complexity defines is length and breadth. Our science confirms objectively from whence the bridge began and our theology and faith intuit its seemingly final destination. Some would call that heaven. I would agree with that assessment. Some would call it a successive emergent and hyper-conscious state – the Omega Point. I would also concur.

As Teilhard's Arrow ever advances, let us connect these seemingly disparate extremes and conclude that a great bridge has crossed an expanse and continues to extend itself into the unknown and unseen. Let the scientists and ardent believers of faith join hands and cross this bridge together.

CHAPTER ELEVEN

DEPARTURE FROM THIS STRANGE WORLD

ALBERT EINSTEIN AND HIS FRIEND MICHELLE BESSO IN 1955

Upon Michelle Besso's death in 1955, Einstein wrote a letter of condolence to the Besso family—less than a month before his own death—which contained the following quote "Now he has departed from this strange world a little ahead of me. That signifies nothing. For those of us who believe in physics, the distinction between past, present and future is only a stubbornly persistent illusion."

I have always been intrigued by Einstein's observation as he approached the end of his life. It seemed odd to me that the renowned physicist known for the precise clarity of his equations would characterize the very nature of time as being an illusion, albeit, stubborn and persistent. It might have been that the great physicist had finally come to the conclusion that the emerging field of quantum mechanics could never be reconciled with his unique contribution to mankind – the Theory of Relativity. Notwithstanding the accuracy and brilliance of Einstein's calculations which were premised on the finite limitation of the speed of light, quantum mechanics and in particular, quantum entanglement, proved precisely the opposite. Einstein was never able to disprove that atomic particles moved across time and space at an infinite velocity. His description of this phenomena was most telling, i.e., "spooky action at a distance".

Quantum entanglement is a process whereby the change in the condition of one particle directly effects the condition of a second particle even if the particles are separated by vast distances. For example, particles can be measured as having properties such as spin along a set axis. If a particle is found to have a clockwise spin at a certain axis, a corresponding particle will be found to have the opposite property, i.e., a counterclockwise spin. Interestingly, if you physically change the state of the first particle to a counterclockwise spin, the corresponding particle demonstrates precisely the opposite spin – clockwise. These phenomena whereby information is exchanged is referred to as "quantum entanglement". The fascinating conundrum commences when you separate the above particles by a distance greater than the time it takes a particle to travel at the speed of light. For example, the speed of light travels at the rate of 299,792 kilometers per second. Let's assume that we separate the above two particles by 1,498,960 kilometers. Even at the speed of light, it would take 5 seconds for particle 1 to travel the distance to particle 2. Now here is where it gets "spooky" as Einstein begrudgingly admitted. The information exchanged between entangled particles proceeds at a velocity in excess of the speed of light. Assuming our two test particles are separated by 1,498,960 kilometers, the change in spin is registered instantaneously. This phenomenon continues notwithstanding an infinite increase in the distance between particles. Theoretically, quantum entanglement can transfer information between a particle on Earth and the very edge of the universe in zero time. Now, how could that be possible?

In 1935, Einstein in collaboration with Boris Podolsky and Nathan Rosen attempted to explain the above phenomena which later became known as the EPR paradox. Ultimately, Einstein was unable to unravel the predictions of Quantum Entanglement and concluded that the theory itself must somehow be incomplete. In the 82 years which has elapsed since Einstein's articulation of the EPR paradox, the violation of the speed of light limitation has been confirmed thousands of times and within different experimental contexts. To this day, the greatest challenge presented in physics (and awaiting the Nobel Prize) is to reconcile Einstein Theory of Relativity with Quantum Entanglement. No one has ever come close to solving this fundamental question.

It is my belief that the inquiry which seeks to reconcile the two theories is a false question and our discussion of Intelligent Process presents the best paradigm for understanding this apparent disparity. One of my core principles is that phenomena is observed differently at various points along the developmental continuum in that emergent states cannot be predicted at any single point. The notion that reality itself is affected by the relative position of an observer on Teilhard's Arrow is brilliantly described in the fictional novella *Flatland: A*

Romance of Many Dimensions by Edwin A. Abbott (1884). The story describes a universe which is occupied exclusively by two-dimensional creatures. In this "Flatland Universe" all physical laws are constrained by two possibilities – left and right. The concept of three dimensions is non-existent and therefore objects have no depth or ability to move up or down. Within this linear dimension, the "Flatliners" observe a strange phenomenon which cannot be explained utilizing the physics of two dimensions. Periodically, there is a juxtaposition of two universes when a three-dimensional sphere interacts with the Flatline universe. As the Flatliners cannot observe – yet alone conceive – of the concept of three dimensions, the sphere appears to them as a circle which appears to emerge and vanish in two-dimensional space. The Flatliner Physicists are baffled by this phenomenon which they characterize as an anomaly.

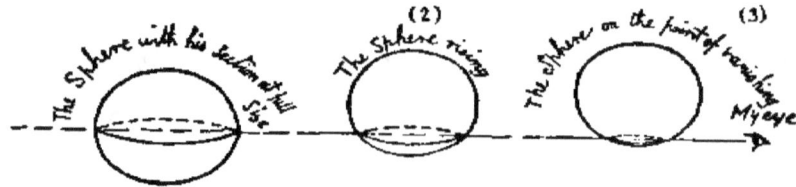

MOONRISE OVER FLATLAND

For purposes of this work, I will refer to the anomalies observed at the point of intersection of two realities or universes, to wit, our existing universe and a successive emergent reality which coexists, as an Emergent Border Phenomenon or "EBP". The above spheres constitute an EBP as viewed by the physicists in Flatland. Similarly, the apparent disconnect between Relativity theory and Quantum Entanglement is an example of EBP as viewed by our physicists.

Applying EBP to the seeming disparity between the macro and quantum states, I would posit that Einstein's speed of light restriction applies to physical reality at our position on the arrow of complexity. From our vantage point as human beings at our current level of neurological function, our world, is fragmented into separate buckets which we refer to as time, space, synthesis, and entropy. Applying what we observe at the macro level, nothing can travel faster than the speed of light and based on such we have extrapolated that the very beginning of the universe proceeded from an enormous explosion 13.8 billion years ago.

That said, the phenomena of infinite speeds observed in quantum entanglement is evidence of physical properties at an emergent state beyond our specific position on Teilhard's arrow and proof positive of EBP. It suggests that there are physical principles contrary to known scientific laws which exist contemporaneously with us. Simply put, quantum entanglement is emblematic of a greater source of complexification which exists in a higher-level emergent dimension beyond our limited plane of awareness and observation. Quantum Entanglement and Quantum Mechanics do not contradict Einstein's equations, they proceed from a point of complexification beyond the limitations of known physics. The EPR Paradox advanced by Einstein is simply an EBP of the same ilk as the "sphere sighting" in Flatland.[xl] Citing again from Teilhard, "If, on the other hand, Omega is, as we have admitted, *already in existence and operative* at the very core of the thinking mass, then it would seem inevitable that its existence should be manifested to us here and now through some traces." (Emphasis in original). I would thus posit that Quantum Entanglement and similar observations which may be noted in the coming years are evidence of the "traces" of Omega theorized by Teilhard. Not only can we intuit the existence of complexification beyond our current state, we can scientifically extrapolate these traces through physical observations which cannot be reconciled with scientific phenomena. EBP's as postulated in this work are direct evidence of the "traces" described by Teilhard. Traces of Omega can thus be verified through physics itself.

All of this is to say that Einstein is correct that "for those of us who believe in physics, the distinction between past, present and future is only a stubbornly persistent illusion." Reality at one point along Teilhard's arrow cannot anticipate or predict the properties of the next emergent state, although we can discern certain "traces". Our notion of the separation between time and space may very well be an illusion from the vantage point of a successive state, i.e., Teilhard's thinking sphere or noosphere or even Omega itself. However, being an illusion in the grand scheme of Teilhard's Arrow does not make it any less real to us. It is only through the discernment of traces, and the driving force of complexity itself, can we gain the perspective necessary to validate the continuing force of Intelligent Process and the true expanse of our spiritual and physical journey.

CHAPTER TWELVE

EMERGENT SPECULATIONS

In this chapter, I will discuss what many would consider to be pure speculation, to wit, a description of what characteristics would inhabit the emergent reality beyond on our current position on Teilhard's arrow. Of course, the very nature of emergent phenomena is that the whole is greater than the sum of its parts and that an observer at an inferior level cannot imagine yet alone conceive of what the next emergent state might entail. That said, here are a few points to consider.

1. Teilhard's Noosphere as Pure Thought – Teilhard describes the emergent level beyond our reality as constituting a thinking mass or

sphere. The concept of "thought" is described as the summit of mankind's intelligence and the manifestation of matter's ability to contemplate itself. So, what would existence in the noosphere be like? Again, engaging in pure speculation, I would envision that all human intelligence would merge into a singular thinking sphere which would have no finite dimensions. The combined memories of mankind would be integrated into a super intelligence and sentience which would be qualitatively distinct from the individual and solitary existence of each one of us. The intellectual power of mankind would expand exponentially to reflect unbridled potential.

2. Departure from Illusions- Einstein described our geographic and tactile references as being illusions with an emphasis placed on concepts such as time and space. A noospheric intelligence which not exist at any one point in time and space would be limitless. An interesting way to contemplate this emergent state is to consider your own thought process. Does it occupy a particular coordinate in time and space? We can imagine ourselves to be anywhere in time and space. Similarly, we have no tactile comprehension of time itself with the exception of crude measuring devices such as clocks. As a 63-year-old man, I cannot fathom what it means to have lived

this length of time or understand in any meaningful way, the passage of the remaining years still to come. I have a sense of the passage of a single year but it is no different than a decade. Most individuals at the end of their life will describe the passage of decades as but a brief moment which cannot be recalled in any concrete way with the exception of time coordinates imposed by calendars and clocks. These observations are consistent with the conclusions of Aristotle who viewed reality as an illusion.[xli] Underpinning what we experience as forms and substance is an unseen world of pure forms and thought. Such speculation going back thousands of years might be right on the money. Similarly, the Greek philosopher Zeno, proposed his famous paradox to negate the very concept of motion. Simply put, Zeno suggests that to get from point A to point B (say one foot apart), requires the participant to cover half the distance. However, to cover half of the distance, similarly requires the participant to cross one half of that distance – and so on. The paradox is obvious – rational thought mandates that motion is impossible and an illusion because an infinite number of intermediate points negates the ability to get anywhere.

The very nature of motion, therefore, may be illusory and abandoned within the emergent reality of the noosphere.

3. Coexistence of Physical Laws – Critically, the nature of an emergent state does not negate the physical properties of the prior state. For example, hydrogen and oxygen atoms retain their separate existence and yet exhibit emergent characteristics in the form of water. Similarly, individual neurons retain their organic characteristics and yet exhibit emergent characteristic within the context of a thinking brain. In the universe of the emergent reality which follows our own, the noosphere would not be subject to any tactile limitation of time, space, or motion. This thinking sphere would be all knowing and unbounded. Concepts of distance, time, space, relativity, quantum mechanics would not be applicable as these separate buckets would only be experienced at the inferior level prior to the emergent state. That said, the noosphere could not exist without the remedial building blocks which precede it and of which combine to generate the successor emergent state.

The most fascinating speculative question of all is what emergent state exists beyond the noosphere? According to Teilhard, the "sidereal force" underlies the propensity of matter to become increasingly concentrated over time. From the emergence of elementary particles to the first atoms, molecules, inanimate matter, organic life, man and consciousness, and the advent of the noosphere, or thinking mass, it is the sidereal force postulated by Teilhard which constitutes a common element of every successive emergent state. Consequently, the only guiding principle available to contemplate the emergent state beyond the noosphere is to recognize that it would be characterized by an even greater level of countrification. Much in the same manner as Teilhard concluded that thought represented matter's ability to reflect upon itself, in contemplating the emergent state beyond the noosphere, I would anticipate that the noosphere would become increasingly centrified and manifest an ability to reflect upon itself akin to matter and the emergence of thought and consciousness. I would further theorize that a super concentrated consciousness beyond the collective consciousness of the noosphere would emerge that would represent a whole greater than the sum of its parts. For lack of a better term, this hyper-consciousness could very well be the Omega Point described by Teilhard.

I would also theorize that this hyper-noosphere would be able to exist independently of its predecessor states which is of critical importance. As previously discussed, each emergent state exists as a layer preceded by a more primitive forms rising to ever greater levels of complexity. As human consciousness is based upon lower levels of organic states, i.e., neurons and the sub-elements of cellular production – the destruction or cessation of organic functioning at a lower level would invariably affect the functioning manifested at greater levels of complexity. Simply put, the death of a human being would kill the underlying neurons and consciousness itself. If the noosphere is based on the prior existence of functioning human brains, it would be evidenced by a rise and fall of consciousness dependent upon the life and death of human beings existing at any one point. It might appear as a great sea of consciousness with the tides representing the rising and falling human brains existing during respective lifetimes. Query – if humanity were to cease to exist, the noosphere would die.

However, the hyper-noospheric state would exist beyond the bounds of prior emergent states and would not be dependent upon prior existing organic forms to maintain its existence and hyper-consciousness. It would be an all-encompassing consciousness which exists beyond matter, space, and time. What better description of this emergent state as Omega or God?

Most interesting, since Omega would exist outside our human understanding of time and space, it would precede all inorganic and organic forms which preceded it as the very notion of time and space would have no meaning. That said, one could conclude from a strictly human tactile perspective that Omega, although emerging at a successive point after the Big Bang, has skipped to the head of the line of creation and would be viewed as a consciousness that pervaded the universe before the very beginning of time and space. So, what came later would become first and the ending would be deemed the beginning of all things.

The final thought I wish to explore is what I refer to as "contemporaneous state/emergence". As discussed previously, the very concept of multiple realities or states along Teilhard's Arrow presupposes a nested hierarchy of increasingly complex structures wherein each state is defined by a new emergent status which is not predictable from the perspective of the prior state. For example, hydrogen and oxygen at the atomic level could be classified as State X and the integrative effect of H_2O as water at State Y. Critically, State X and Y exist simultaneously notwithstanding the inability of a hypothetical observer at State X to predict or even fathom the nature of reality at State Y. In the same fashion, the individual cellular function of a neuron at State X would have defined characteristics and organic properties. However, it is the integrative effect of the organic neural net which leads to the emergence of consciousness. An observer as State X would not be able to envision the emergence of consciousness as a byproduct of the collective function of neurons at State Y.

However, neurons would remain individually poised as functioning units at State X while they interact at a higher level of complexity at the next emergent level State Y.

This same modality would apply to our individual consciousness at our current State X. All of us have our own sense of identity and consciousness. The contemporaneous state/emergence construct would predict that our consciousness is also existing at a higher level in combination with all other existing individual consciousness at State Y which is Teilhard's noosphere. From our perspective at State X as individual thinking entities, we have no concept of the integrative effect of all thinking entities contemporaneously existing at State Y which would manifest characteristics greater than the sum of all individual minds as Teilhard's "thinking mass" or the noosphere. We would thus exist at two points contemporaneously.

CHAPTER THIRTEEN

EMERGENT BORDER PHENOMENA AND THE QUANTUM/RELATIVITY RIDDLE

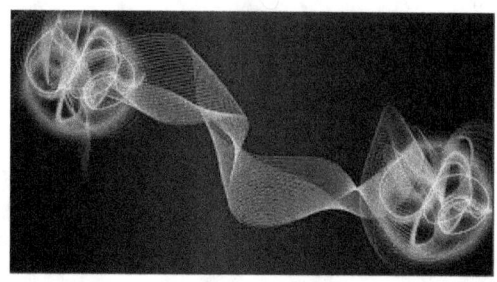

As previously discussed, the concept of Emergent Border Phenomena (or EBP) relates to the observation's attendant to the intersection or border area between an existing State X and the successor emergent state, i.e., State Y. Very much the way the physicists in Flatland could not understand the phenomena of a three-dimension sphere intersecting in their two-dimensional space, our physicists are confronted with the same quandary as it relates to the intersection of the micro world of quantum mechanics and the macro world of our human reality as embraced in Einstein's Theory of Relativity.

I would posit that the strange world of quantum mechanics provides direct evidence and a glimpse of the physical properties of the emergent state beyond our macro view of reality as human beings. The best way to view the respective properties of the quantum and macro is to categorize them as representing distinct states or universes. It is only when we attempt to measure a particle in its quantum state that the so-called wave function (Copenhagen Interpretation) collapses, and the realities appear to break apart from each other and become distinct.

The quantum world as observed by physicists prior to measurement exhibits the following characteristics:

1. Universality of Particles – All particles appear to exist at all possible points at one time which is impossible in the macro world. Thus, the Heisenberg Uncertainty Principle provides that the precise location of a particle can never be predicted, only the probability that it might exist at a particular point. Therefore, the quantum world of particles being located at all points simultaneously is likely the best description of the emergent state beyond our known reality.

2. EBP Observed within context of Measurement- An EBP is always observed when a particle is measured. At that precise moment, the particle assumes a particular place in space which cannot become

uncertain within the confines of a strictly quantum state. The seeming disconnect between the uncertainty at the quantum level and certainty at the macro level is not a disconnect which needs to be spanned or a riddle which needs to be solved. The world and properties are distinct. It is only at the point of intersection that the EBP is observed which is the best explanation of what otherwise appears to be irreconcilable principles or states.

3. Quantum Entanglement- As previously discussed, particles in their quantum state exchange information at a velocity faster than the speed of light. Once again, this entangled interconnected state which likely commenced upon emergence of particles after the Big Bang best describes the emergent state beyond our physical universe. It cannot be reconciled with relativity theory as the realities are distinct. Einstein's inability to explain this process is again consistent with his observation of an EBP.

4. Transfer of particles between distances in zero time – Quantum observations provide that particles appear to appear and disappear in zero space and time. A particle can move from position x to y in zero time which is impossible in our macro world. Such is evidence

of the distinction between the respective realties and the intellectual conundrum presented yet another example of an EBP.

5. Many Words Interpretation – This is considered to be an alternative to the Copenhagen theory which provides for the collapse of the wave function. Supposedly, when the quantum become actual, the universes split and multiple realties are manifested in multiple universes. This seem absurd to me as the collapse of the wave function best describes the intersection of our physical reality and the emergent reality which is observed as an EBP.

13.1 The Double Slit Experiment

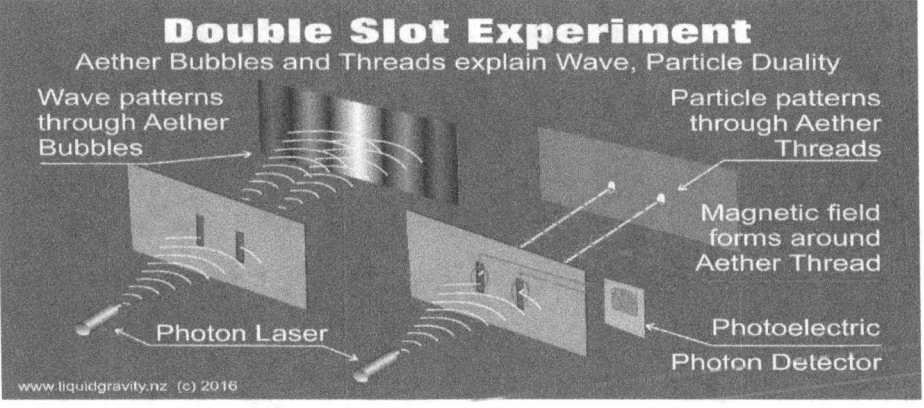

The great physicist Richard Feyman characterized the results of the double slit experiment as the greatest unanswered question presented in physics and quantum mechanics in particular. The experiment consists of a wall with two vertical slits on either side. A projection device then shoots a variety of particles through one of the slits including photons, electrons, atoms, and simple molecules to obverse the effect of the particles striking a photo sensitive plate or screen several feet from the slits. The resulting image consists of a series of dots representing the photons or other particles striking the back plate. A composite image of the aggregate of all dots forms what physicists refer to as an "interference pattern" which consists of a series of white and dark stripes.

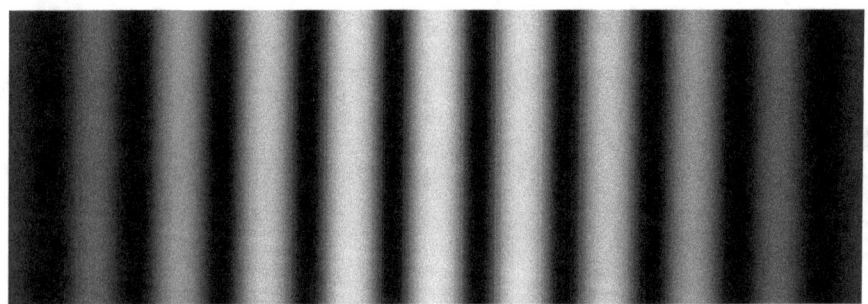

INTERFERENCE PATTERN

To better understand the basis for the interference pattern it is essential to view light particles (photons) or any particles as a duality consisting of individual packets of energy but also a wave. This wave particle duality is an essential principle of physics and has been observed since the days of Isaac Newton. However, it is a baffling notion that a single photon, for example, can exist simultaneously as a discrete particle and also a wave. The interference pattern above occurs when two waves intersect with each other allowing the tops of each wave to combine and thus form a bright line and the troughs and corresponding peaks cancel each other out and thus form a dark area. The best way to visualize this is to consider ocean waves rippling through the ocean. The combination of two waves will result in the cancellation of certain waves and the amplification of others as seen below.

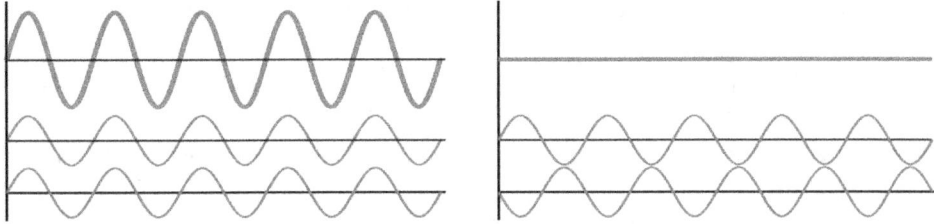

COMBINANT EFFECT OF WAVES

Now, the true "weirdness" of this double slit experiment materializes when scientists are able to project a single photon through one of the slits one by one, let's take the left slit as an example. A series of photons are shot through the left slit and the resulting impact on the plate is observed. Of Couse, we would assume that the impact would result in a series of single particles on the photo sensitive plate in the shape or outline of the vertical slit. However, what appears is an interference pattern which suggests that at least two particles must have been sent through both slits allowing for the interference pattern to emerge, but not so! So, what is the explanation for a single photon *interfering with itself* and forming the multiple wave combinations of an interference pattern? Quantum Mechanics theorizes that the single photon is present at all locations in the universe – not just the vicinity of the wall and two slits! This sounds mind boggling, but the underlying mathematics of this explanation has been verified thousands of times without error! So, what is going on? The photon is physically present as a wave at all locations simultaneously. Therefore, as the single photon enters the left slit, it is also simultaneously entering the right slit and every space in between, up and down, and everywhere in the universe. The resulting effect is the combination of waves (yes – plural and not singular) which cancel out or amplify in the familiar interference patterns! What is even more mind boggling is that when a particular hit the screen behind the left slit (say particle 2), it has no idea where particle 3 or 50,000 for that matter will land. That said, the particles somehow know where to deposit themselves on the screen suggesting a prior knowledge of future events! Again, once all the particles are registered (50,000 in our hypothetical), the combined effect of all the particles form an interference pattern. To date, no viable theory has been advanced to solve this riddle, at least until now.

This author would suggest that the emergent state beyond our conscious perspective consists of all energy, whether photons or sub-atomic particles as one unified field. This would explain the concept of non-locality and entanglement where one particle can affect another particle notwithstanding the separation of vast distances as detailed earlier in this work. It would also explain the phenomena of particles which seem to appear and disappear out of nowhere and to cross distances in zero time. Clearly, the quantum world phenomena are best explained as constituting a separate emergent state beyond our emergent position at a junior position along Teilhard's arrow. There is no need to attempt to reconcile this quantum emergent state with our own reality as they are conceptually distinct. Einstein's observations embodied in his

Theory of Relativity define the physical laws which exist at our emergent level of existence. The quantum observations define the physical laws which exist at the next successive emergent level, and never shall the twains meet. I believe it is that simple.

Moreover, it is my opinion that the Copenhagen Interpretation whereby the wave function collapses at observation best explains this interface between the two emergent states, the lower level macro world of relativity theory versus the higher-level quantum world of quantum mechanics. As previously discussed, the Copenhagen Interpretation suggests that at the point observation, the quantum wave of potential physical states, for example, our photons being projected successively through double slits, collapses and appears as a single particle or point. The mathematical equations of quantum mechanics can actually predict with close to 100% accuracy where a photon will appear upon observation (collapse) based on probability theory.

It is again baffling to imagine that our human observation results in the collapse of the quantum probability wave and its transformation to a single particle at a fixed location. As above discussed, we know that the individually projected photons upon impact (at observation) form the aggregate of an interference pattern. This suggests several critical conclusions:

1. The observation that a particle is interfering with itself and forming an interference pattern proves that the particle is occupying an infinite number of points in space simultaneously. This suggests that there is a unified field formation which constitutes the senior emergent quantum state that exists contemporaneously with our junior macro emergent state.

2. The actual objectification and of these results as interference patterns is proof positive of a verification of what this author terms a Quantum Border Effect.

3. It is the very process of human observation which causes an interface between the two successive emergent states (senior quantum emergent state and junior macro emergent state) to occur and result in the verification of interference patterns. Therefore, human observation does collapse the quantum wave causing an interface between the two successive emergent states. The senior emergent quantum state at collapse is observable in our universe (lower nested level) as an interference pattern

which documents the existence of a quantum border effect as postulated by this author.

The above conclusions would further lead this author to speculate that our consciousness exists at two emergent states contemporaneously, which forms the very basis of the theory presented in this work. Our minds at the unconscious level consist of pure thought which likely exists at the pure quantum level, or noosphere. Pure thought would exist as a quantum field or whole and have no discrete particulate properties. However, the application of human sensory organs, i.e., sight, sound, etc., would collapse the pure thought or noospheric field, into discrete particles subject to human observation. Essentially, the existence of human consciousness as expressed through our sensory organs, would effectuate the collapse of the quantum wave (or noosphere) and result in our perceived reality at the macro emergent level of our existence as discrete particles. Taking this concept further, the human mind would constitute the interface point between the two emergent states (lower macro and upper quantum), collapse the noospheric wave or field, and result in our perception of reality. Therefore, everything we perceive as human beings is nothing more than the transformation of a wave to discrete particles. Accordingly, our sense of reality is not representative of reality, but a limited rendering of a higher reality played out at a lower collapsed particulate universe which only exists as a consequence of our consciousness itself. Taking this further, our very sense of finite space and movement from one physical point to the next would be an illusion. The same would apply to our perception of time.

It is important to expand our discussion as to whether this illusion permeates what we consider to be our reality. Assuming that the human mind collapses the wave, function and reduces matter and energy to discrete points versus an unbroken continuity, our very notion of motion and time may constitute what Einstein referred to as a "stubborn persistent illusion".

ZENO OF ALEA

We previously discussed the paradox posited by the ancient Greek philosopher Zeno of Elea (490-430 BC). His underlying assumption was that since the movement of a tortoise or arrow, for that matter, required an infinite number of intermediate steps, one could never get from Point A to Point B. The same paradox would apply to our very notion of time itself. Take, for instance, the elapsed time for the hands of an antique grandfather clock to move from 11:59 AM to midnight. Sixty seconds can also be broken down into a subset of infinite intermediate time points. Thus, the very passage of time would also appear to be illusory. This analysis is critical as it would suggest that everything we perceive as reality is illusory. We never really move from one point to the next even if we experience the passage of distance and time. Such conclusion is not novel as it was first suggested by Aristotle (384-322 BC) in his seminal work, *Physics*. In all likelihood, we inhabit a continuous and unified spacetime manifold which only appears to allow for motion and time passage. We might as well be sound asleep in a cosmic bed living out our life in a dream. I would equate this dream component as our emergent existence within the noosphere and illusory observations of motion and time which form the basis of our personal memory to reside.[xlii]

The counterpoint to this argument is that modern mathematics has solved Zeno's Paradox and that the real illusion is not motion or time, but Zeno's Paradox itself. However, a minority of mathematicians (of which I ascribe) argue that Zeno's Paradox has not been solved by Calculus. As a leading opponent state:

"One common reply is that Zeno misunderstood the nature of infinity. Modern mathematics, it is said, has shown that the infinite sequence that Zeno generates do have a finite sym. In particular to take the Racecourse example, the sequence ½ + ¼ + 1/8 + 1/16... is equal to 1. This reply, however, misunderstands what modern mathematics has shown. Mathematicians do use sequences such as ½ + ¼ + 1/8 + 1/16... but they say that they have a *limit* of, or *tend* ı (emphasis in original). That is, we can get nearer towards 1 by adding on more numbers to the sequence, but not actually arrive at 1 – this would be impossible because we are considering a finite sequence. So far from proving an argument against Zeno, mathematics is actually agreeing with him!"[xliii]

This phenomenon, including the physical laws opined by Einstein which govern same, are only applicable to our lower nested emergent state.

Thus, our minds would occupy contemporaneously, the higher quantum state of the noospheric field (pure thought) and our sense of consciousness at the lower macro state of individual thought (reflection). It may be correct that we are living contemporaneously in successive emergent states.

I would further posit that at the level of thought, there are two levels suggested, the thought byproduct of consciousness and wave collapse and pure thought. As our consciousness collapses the quantum wave, our brain encapsulates observations based on discrete points which form the basis of our memory. Although this perception is adulterated and not indicative of the higher emergent quantum state, we perceive it as real and form memories of same. Our lives consist of memories which are also embodied in the very language and symbols we use to store information. This information is then maintained as human thought within the personal morphic field previously discussed. However, the emergent state beyond the personal morphic field – the collective morphic field previously discussed, would constitute the realm of pure thought or higher-level area within the noosphere.

13.2 The Q Manifold and Anthropic Principle Revisited

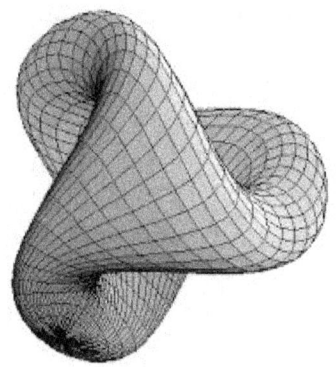

The anthropic principle postulates that our universe exists because we are here to observe it. This theory is troubling to physicists as it seems to suggest that our universe and reality are not based on objective natural laws but some external deist cause. A typical example of this playing out is the likelihood that the cosmological constant would be arrived at by a natural process. The cosmological constant is an equation which provides for a counter balance against gravity. This calculation is the most finely tuned equation. The statistical chance of the cosmological constant being produced by random process is 10 to the negative 122 or 10 with 122 zeros (one chance in a trillion trillion trillion trillion trillion trillion! This infinitesimal probability so often cited by intelligent design proponents as proof positive of the existence of a designer or God.

I look at this situation differently. Going back to the basic concepts of quantum mechanics – double slit and Heisenberg Uncertainty Principle, particles exists at all points in the universe simultaneously prior to human observation. Thus, the probability of an event occurring whether it is 1 in 5 or 1 in 10 to the negative 122 is played out at inception in a singular manifold which contains all possible combinations. I refer to this as the Quantum Total Probability Manifold or the Q Manifold. Every particle location exists simultaneously in the Q Manifold which would include the underlying combinations and physical laws governing same. Consequently, at the point wherein pure energy after of Big Bang condensed to form the first particles, the Q Manifold would have come into existence which would have included the cosmological constant itself resulting in the formation of the universe we inhabit and observe. This must be true unless the probability of a result occurring is "0". Even the chances of the cosmological constant occurring by a natural process is great than zero.

Another way to understand the Q Manifold is to consider a typical lottery game. Let us say there is a 1 on 100 Million chance of winning each time we purchase a single ticket. Therefore, our chance of winning would ultimately result in the big money after 100 Million attempts. However, from the standpoint of the lottery ticket manufacturers, if 100 Million tickets are printed and purchased, a winner must immediately exist. Consider the total number of tickets printed and sold as the Q Manifold. At inception of the Q Manifold's existence, a winning lottery ticket must exist. Now, from a cosmological perspective, the same concept would apply. If there is a 1 to the negative 122 chance of the cosmological constant arising from a random process, the end result would be almost impossible to occur by a random process. Thus, the intelligent designers would proclaim that God must have originated the cosmic constant and everything else at creation. However, if one views all potential configurations of particles as occurring simultaneously as quantum mechanics predicts, the Q Manifold would have arisen at the moment all particles originated at the condensation after the Big Bang. Within the Q Manifold, the outcome of 1 to the negative 122 would have originated at the inception of the Q Manifold's existence. Thus, the probability of all particles and configurations would occur at inception is analogous to all lottery tickets being printed and sold. Our particular universe which gives rise to us and as a consequence we observe, is the winning ticket.

The intellectual obstacle to surmount is to avoid viewing the infinitesimal chances of our universe arising as a sequential or arithmetic result as a linear path will never get you to the final destination, at least without the passage of many trillions of trillions of trillions. However, if all possibilities exist simultaneously at inception, the right answer exists at a point within the Q Manifold. The infinitesimal sector of the Q Manifold that we inhabit is erroneously referred to as a separate universe within multiuniverses. The more rational view, in my opinion, is that our universe simply exists within a finite location in the Q Manifold.

Critical to the notion of "Intelligent Process" versus "Intelligent Design" is that a God or sentient intelligence exists now, but emerged as a result of complexification. This is to be distinguished from the traditional "Intelligent Design" approach that God preceded all things.

13.3   Dark Energy & Matters Revisited

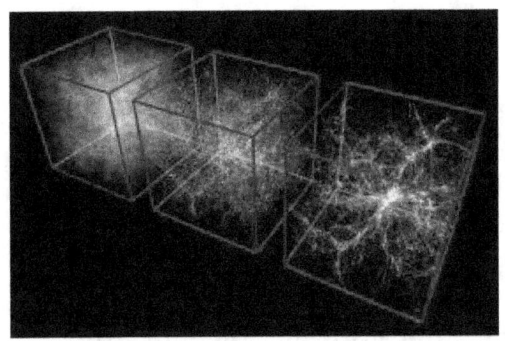

GEOMETRIC VIEW OF DARK MATTER & DARK ENERGY

Another possibility is to consider the manifold as being an entirely different structure. In this further speculation, we might consider space to be the exterior level or shell of the manifold wherein would be what we currently refer to as dark energy or dark matter. Cosmologists and physicists, particularly those who ascribe to the inflation theory, already consider the expansion of the universe as driven by the increasing expansion of space itself. Therefore, when cosmologists discovered that the most distant objects of the universe were receding at ever greater speeds, the typical description which focuses of the "speed of the objects receding" is wholly inaccurate. The objects are not receding but simply positioned upon the underlying space which is rapidly expanding in size. A typical metaphor is to consider space to be the surface of a balloon which is being inflated. Points affixed along the surface of the balloon will appear to move apart as the volume of the balloon increases. The same situation applies to space itself. The operative question, therefore, is -- what is the substance or force within the balloon, or in our case, the manifold, analogous to the expansive force of oxygen? Dark energy and dark matter would seem to be the perfect candidate.

Moreover, if one were to consider our perception of space is merely the exterior surface of the manifold, our visible universe within this shell layer of space would constitute a very small percentage of the total volume of the manifold. This seems to comport with recent findings that dark energy and dark matter account for approximately 96% of all matter in the universe. We and all of the observable objects in the visible universe occupy a mere 4%.

I believe it is fruitful to consider basic physical forces as a byproduct of geometry. Einstein, of course, changed the Newtonian view of gravity wherein objects exerted a pull on each other. Physicists now understand that gravity is simply the result of the curvature of space caused by the effect of matter. The greater the matter in a given area, the greater the curvature of space. Objects thus fall into or assume a path along the indentation caused by the curvature of space. Interestingly, this concept is virtually unknown to the general public.

Taking this geometric approach further, we could consider our universe as simply occupying a particular position along the manifold which continues to accelerate its expansion based on the underlying effects of dark energy and dark matter. Consequently, it is better to consider the concept of the multiverse as not constituting separate universes but different locations or positions along the manifold. I would posit that different configurations of matter and underlying realties are entirely based on a discrete location along the manifold. Therefore, I would suggest eliminating the term "multiverse" and substituting the term "multi-manifold coordinates". Now, the X and Y axis of any coordinate along the manifold would represent the conjunction of the physical properties present at the discrete location on the surface of the "space shell" (X axis), and the concentration of matter as the Y axis. Again, visualizing space as the exterior or shell of the manifold, the concentration of matter at a particular location would cause the surface of the manifold to curve as Einstein predicts. Therefore, if one were to view the surface of the manifold shell, it would appear to be cratered with the depth of each crater equating to the concentration of matter and resulting indentation. Continuing our application of geometric visualization, any point along the X, Y manifold would be experiencing two independent effects, to wit, the outward expansion of the manifold shell (which we view as the accelerating expansion of the universe), and the force of gravity which is the depth of the curvature of the manifold crater, so to speak.

Continuing down this line of inquiry, it may be possible that the rate of inflation varies depending on the location of the X, Y manifold. Assuming that the rate of expansion is greater at some points along the manifold in relation to others, the geometric visualization would be that a point at X,Y manifold 1, for example, would have a different rate of expansion of the manifold in tandem with the curvature of space caused by mass, then another point along the manifold, say X,Y manifold 2.[xliv] If this is correct, then what we refer to as the cosmological constant might simply be a geometric location along the manifold which represents the self-tuned ratio of expansion to gravity resulting in the formation of matter in our perceived reality.

The final element in this conjecture is that the basic subatomic particles captured by our quantum mechanics theory would be the only constant along the manifold. Depending on the location of the elementary particles at a particular manifold X, Y coordinate, atoms would be able to combine to form the objects of our observable universe. At other points along the manifold, matter would either not be able to form or would congeal in an entirely different process.

Continuing the employment of the manifold as our visual tool, the "space shell" might be viewed as continents on the surface of the manifold assuming varying shapes and forms with inlets, peninsula, etc. Our so-called universe, would simply be a location in some inlet along the manifold itself, wherein the ratio between expansion of the manifold and corresponding curvature of same, is so finely tuned that our reality came into existence.

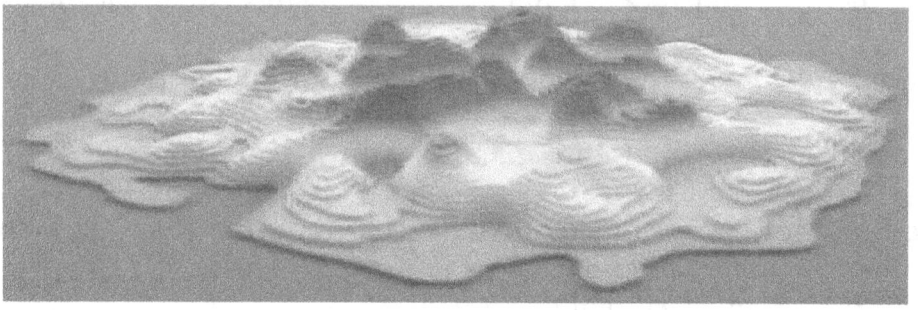

3D VISUALIZATION OF COORDINATES ALONG THE X, Y MANIFOLD

CHAPTER FOURTEEN

CONSCIOUSNESS ALONG TEILHARD'S ARROW

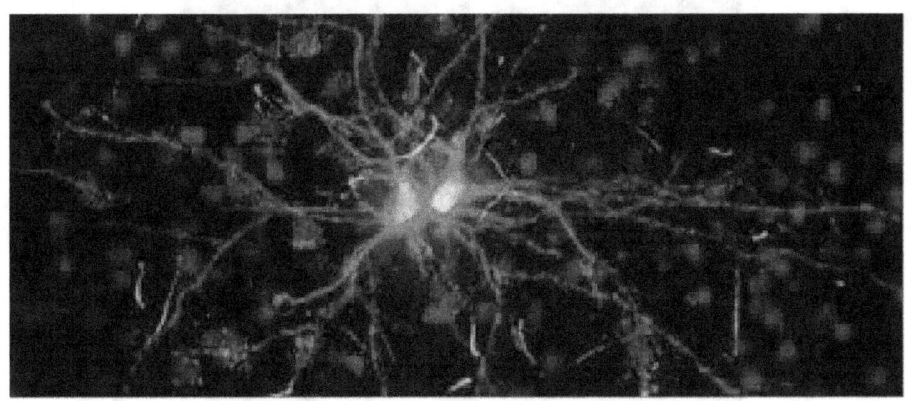

HUMAN NEURONS

Although science has made great strides in the 21st Century, neurology is still in its infancy and the most basic understand of memory and consciousness has yet to emerge. Teilhard's concept of "thought" as an element of complexity may provide an integrative construct for better understanding these fundamental processes.

My inquiry begins with the scientific studies of W. McDougall, a psychologist at Harvard University in 1920. He was interested in testing a basic premise of Charles Darwin that natural selection was the driving force of evolution.

WILLIAM MCDOUGALL

McDougall sought to determine whether the Theory of Acquired Characteristics advanced by Jean-Baptiste Lamarck (1744-1829) offered a better explanation of the evolutionary process. Lamarckism asserts that organisms pass on the acquired traits to their offspring and is thus the compelling force of evolution. His theory became widely discredited in the wake of Darwin's publication of On the Origin of the Species in 1859 and the adoption of natural selection as the operative principle of evolution by the scientific community. That said, interest in Lamarck's theories persisted as embodied by William McDougall's fascinating research in the 1920's.

McDougall graduated from Owen Colleges and St. John's College in Cambridge. After teaching in England, he moved to Harvard University where he served as a professor of psychology. He is best known for devising a test in laboratory mice to determine whether the Lamarckian Theory of Acquired Characteristics was valid. His ingenious test challenged a unique strain of laboratory white mice to escape from a tank of water by swimming to one of two possible exits. The wrong gateway was brightly illuminated whereas the second correct exit was dark. Mice entering the illuminated gateway also received an electric shock. The mice were immersed in the tank and subjected to repeated trials to determine their rate of learning. The number of errors made by a mouse before it learned how to exit the tank provided an objective measure of the rate of learning. McDougall noted that "Some of the rats required as many as 330 immersions, involving approximately half that number of shocks, before they learnt to avoid the bright gateway. The process of learning was in all cases on which suddenly reached a critical point.... After attaining that point, no animal made the error of again taking the bright gangway, or only in very rare instances."[xlv] McDougall conducted this experiment of 32 generations of mice for a period of 15 years. In accordance with Lamarckian theory, there was a marked tendency of mice in subsequent generations to learn faster. The results were pronounced wherein there as an average of 200 wrong exits taken by the mice before learning the task at generation one. This declined to a mere 21 errors at generation 21 and continued thus until the termination of the experiment.[xlvi]

The mainstream scientific community adhering to Darwin's paradigm was skeptical and McDougall's research was repeated by F.A.E. Crew, a scientist based in Edinburgh. He studied over 18 generations of mice and his results at first seemed to dispute McDougall's conclusions. He initially determined that there was no difference in the rate of learning between the trained and untrained mice. Critically, he discovered that his test mice were able of learn to correct exit faster than McDougall's mice. In fact, the mice during the initial generations of the test achieved the same rate of learning as McDougall's mice after 30 generations of training. To add a further element of astonishment to this equation – the trained and untrained mice acquired the skill at the same rate and some without the administration of an initial shock. The inexplicable conclusion of this research was that the learned behavior was more pronounced in both populations (trained and untrained) once the skill had been successfully acquired. Simply put – it became easier and easier for successive generations to learn the new skill. Although not discussed in the literature at this time, the results seemed to suggest that there was some sort of collective memory process involved independent of any particular generation of mice.

RUPERT SHELDRAKE

Enter on to the scene – British biologist Rupert Sheldrake who set forth a theory which explained the phenomena observed by McDougall and F.A.E Crew. Sheldrake (born in 1942) was educated at Clare College where he became a fellow. He proposed a unique and astounding theory known as "morphic resonance" which advances that memory is inherent in nature and is actually stored and transmitted from fields of information residing outside the brain.[xlvii] Although Sheldrake's conclusions have not been accepted by the mainstream scientific community, they offer an explanation of McDougall and Crew's results. Applying Sheldrake's theory, the learned behavior observed by McDougall required multiple generations to be fully realized by the mice, to wit, 32 generations over 15 years! However, repeated experiments by other researchers documented that once the skill was learned by successive generations of mice, the speed of acquisition decreased for both trained and untrained generations. Crew's results demonstrated that untrained mice learned the behavior in a single generation! According to Sheldrake, once acquired, a collective memory resides in a field information particular to the mice but outside their physical bodies. Memory, therefore, resided outside of the brain and can be collectively accessed by other organisms of the same species. This is truly a ground-breaking hypothesis and offers connectivity to Teilhard's theory and the thesis of this work – Intelligent Process.

Sheldrake's field of information or morphic field is analogous to Teilhard Noosphere which exists as an extrinsic thinking mass or field of awareness outside of human consciousness. I would posit that the subject mice had access to information at two emergent levels, to wit, their individual awareness and contemporaneous existence as part of a Noosphere or collective memory of their species. Thus, the mouse would have the ability to acquire memory and skills within its solitary existence and yet simultaneously access the morphic field or noosphere of all mice. In the same fashion, human beings would have knowledge of their own discrete existence and their primary level and yet contemporaneously access memory and awareness at the collective level of the morphic field for humanity. For those who are visually inclined, I would draw a circle above an individual as the discrete morphic field of that person and history. Experience relating to that person would reside in the individual morphic field, so to speak. To put it in personal terms, this author's individual morphic field would consist of everything that constitutes his personal history and sense of identity. I was born in Manhattan, NY in 1954 to Joe and Sandy Moster. I spent my early childhood in Long Island and moved to the Miami area in 1967. I was educated at Georgetown University and other institutions. I pursued an early career in Theatre and went on to study law. I am happily married (after two failed attempts) and have a young daughter at age 63. My personal memory is rich and nuanced including passions, first loves, experiences of pain and pleasure, and lessons learned over a long period of time. My first dog was Diamond and I am currently at my desk at the Moster Law Firm which is a business I created in 2010 in Lubbock, Texas. All of this is the Charles Moster Individual Morphic Field and exists not in my brain but a field of information outside my physical body.

Concurrently, my Individual Morphic Field is a component or element the Collective Morphic Field of all humanity encompassing all memory from the first generation of humans to now – akin to our white mice. This Collective Morphic Field is best described at Teilhard's Noosphere and is an emergent level of consciousness beyond my Individual Morphic Field. I share this Collective Morphic Field or Noosphere with all humans on Earth and all of my ancestors from the beginning of civilization. Critically, the nucleus of this Collective Morphic Field is Teilhard's Noosphere which possesses an emergent consciousness which is greater than the sum of all the parts – to wit, the billions of humans on earth and even greater number that preceded us. On a personal level, Charles Moster, Individual Morphic Field, circa 1954 to Present, co-exists contemporaneously with the Collective Morphic Field of all humanity and is both a component of that (and all human memory) and synthesized within the centrified Noosphere itself or thinking mass. I thus occupy two states although my consciousness is based on my awareness at my emergent level along Teilhard's Arrow – the Individual Morphic Field.

RAY KURZWEIL

14.1 THE NOOSPHRE AND SINGULARITY

A very interesting question is whether the increased rapidity of learning observed by McDougall and Crew can be correlated with Sheldrake's theory and applied to Teilhard. As you will recall, McDougall and Crew experimentally established that the rate of learning increased as the test subjects (mice) were exposed to conditions over time. Sheldrake attributed this phenomenon to the development of a morphic field of memory and collective access by species. Would not Sheldrake's theory suggest that the collective morphic field of humanity has retained and built upon the memory and experiences of all humans who have ever lived? If so, would we not experience an increase in the rate of learning or knowledge acquisition as a species? Teilhard, of course, would view this collective memory and augmenting knowledge base as the noosphere.

Ray Kurzweil, a renowned computer scientist and inventor postulates that humanity is in fact experiencing an acceleration of knowledge as evidenced by our inventions and new technologies, particularly Artificial Intelligence.[xlviii] The growth of knowledge is thus exponential, according to Kurzweil, and exhibits the accelerating returns of a positive feedback loop. This phenomenon is particularly evident in computer technology which is rapidly advancing beyond the mental capacity of humanity and will exceed the sum total of our knowledge in short order. The ultimate point of convergence as theorized by Kurzweil is known as the "Singularity" and has created quite a stir in popular and scientific discussion.

I agree with Kurzweil to the extent that knowledge is expanding at an exponential rate and would refer to the research previously discussed as evidence of same, i.e., McDougall and Crew, among others. However, I disagree that we are moving as a species toward a singularity as such singularity is already in existence as postulated by Teilhard and in this work. The ultimate singularity is Omega which exists as a sentient force as the successive emergent level beyond the Noosphere. That said, at the level of Collective Morphic Field or memory, we would experience what Kurzweil refers to as an emerging singularity. Such convergence, however, would be limited to our position along Teilhard's arrow and thus compartmentalized. I believe that Kurzweil is incorrect that once the singularity occurs, intelligence will radiate outward from our planet to the universe. Such hyper-intelligence radiated long ago and is already in existence.

14.2 CARL JUNG AND THE COLLECTIVE UNCONSCIOUSNESS

CARL JUNG

The above concepts are also analogous to Carl Jung's theory of the collective unconsciousness and archetypes. Carl Jung (1875-1961) was a

Swiss psychiatrist who coined the phrase "collective unconscious" within the context of his theory of archetypes. A description of this concept was set forth in a lecture he delivered in 1936. "My thesis then, is as follows: In addition to our immediate consciousness, which is of a thoroughly personal nature and which we believe to be the only empirical psyche (even if we tack on the personal unconscious as an appendix) there exists a second psychic system of a collective, universal, and impersonal nature which is identical in all individuals. This collective unconscious does not develop individually but is inherited. It consists of pre-existent forms, the archetypes, which can only become conscious secondarily and which give definite form to certain psychic contents".[xlix]

Jung's description of the collective unconscious is certainly analogous to Sheldrake's morphic field and this author's reference to the Collective Morphic Field. Critically, Jung recognizes a distinction between the "personal consciousness" and the deeper "collective unconsciousness" which is universal in nature and appears to have inherited characteristics.[l] The "personal unconsciousness" is obviously analogous to what this author refers to as the Individual Morphic Field.

At the heart of Jung's theory is his concept of archetypes which constitute structures within the landscape of the collective unconscious. He describes archetypes as follows: "They evidently live and function in the deeper layers of the unconscious, specially in that phylogenetic substratum which I have called the collective unconscious. This localization explains a good deal of their strangeness: they bring into our ephemeral consciousness an unknown psychic life belonging to a remote past. It is the mind of our unknown ancestors, their way of thinking and feeling, their way of experiencing life and the world, gods and men. The existence of these archaic strata is presumably the source of man's belief in reincarnations and memories of "previous experiences". Just as the human body is a museum, so to speak, of its phylogenetic history, so too is the psyche."[li]

In my view, Jung's theory of archetypes refers to the outward structure of the superficial levels of Teilhard's Noosphere. Certain recurring categories or themes which characterize human nature and the knowledge of all things relating to humanity may have defined structure within the Noosphere, similar to the way the cerebrum is distinct from the cerebellum in the brain. These archetypes, however, evidence the structure and order of the Noosphere and not its actual sense of individuation as an emerging thinking mass within the Noosphere much in the same way as the location of the cerebrum or cerebellum is not emblematic of consciousness itself. That said, this recurring structure can be perceived and recognized by individuals at a personal level within the context of philosophical or psychiatric works as above noted. I would categorize these structures as a type of Emerging Border Phenomena previously discussed in this work. Although we can only vaguely detect its presence, it suggests an underlying hidden structure of a successive emergent state beyond our existing position along Teilhard's Arrow.

14.3   SYNCHRONICITIES

SYCHRONICITY ACAUSAL CONNECTIONS

The term "Synchronicity" was also described by Carl Jung as a hidden structure beyond our sense of reality and a series of "acausal connections".[lii] Jung further described the concept of synchronicity as follows: "How are we to recognize acausal combinations of events, since it is obviously impossible to examine all chance happenings for their causality? The answer is that acausal events may be expected most readily where, on closer reflection, a causal connections appears to be inconceivable."[liii] An excellent example of a reported synchronicity appears in this appendix.[liv]

The work of German biologist Paul Kammerer and his theory of seriality is an adjunct to the concept of synchronicity. He theorized that events in our life are connected by "waves of seriality". These acausal connections are perceived by us a coincidences or inexplicable groupings of events. Kammerer believed that these waves of seriality suggested an underlying and hidden structure to reality as also advanced by Carl Jung. Kammerer could often be observed sitting for hours at a park copiously observing and writing down events in an attempt to decipher hidden patterns. Albert Einstein characterized Kammerer's work as "interesting and by no means absurd"[lv].

PAUL KAMMERER

In my view, synchronicities provide a momentary window into the structure of the next emergent point along Teilhard's arrow – albeit the superficial organization of the Noosphere. The brilliant artists, poets, and scientists have an ability to perceive this inner Noospheric structure and to bring forth that information into material existence.

I also believe that other individuals have the ability to access the prior memories of individuals now departed whose memory is part of the collective unconscious or Noosphere, again at a superficial structural level. As McDougall and Crew suggest, memory is not lost but stored in the collective unconscious or Noosphere as Teilhard would term it. Although I am skeptical of many accounts of reincarnation, the academic reports of children reporting past life experiences are literally draw dropping. I believe that such incidents are evidence of an ability to access the actual stored memory of a past life. A fascinating description of these inexplicable recollections are set forth in the book Return to Life: Extraordinary Cases of Children Who Report Past Lives[lvi]

CHAPTER FIFTEEN

CONSCIOUSNESS AND ARTIFICIAL INTELLIGENCE

IBM BLUE GENE/P SUPERCOMPUTER AT ARGONNE NATIONAL LABORATORY

As a child, I was intrigued with the possibility that computers in the near future would become fully conscious. I remember how astounded I was at the New York World's Fair in 1964 to witness fully functioning robots which approximated the movements and behavior of American Presidents and, of course, folks like you and me looking forward to the future of home appliances at the General Electric Exhibit – "It's A Great Big Beautiful Tomorrow". Hollywood amplified my excitement and anticipation of conscious computers in such blockbusters as 2001: A Space Odyssey (1968) which featured the malevolent supercomputer Hal 9000 and Colossus: The Forbin Project (1970), which had the identical intent as Hal but actually pulled off the conquest of mankind (no spoiler warning required).

**HAL 9000 COMPUTER
FROM 2001: A SPACE ODYSSEY**

Notwithstanding my AI enthusiasm, the development of a conscious computer is as elusive in 2018 and it was in the late 20th Century. To be sure, our computers have become ever more powerful, but Hal is nowhere to be found except in Science Fiction.

The first commercial computer, UNIVAC-1, was manufactured in 1951 for use by the U.S. Census Department and performed 1,905 operations per second with the assistance of over 5,000 vacuum tubes.[lvii] Compare this ancient relic with the fastest computer in the world developed by the Chinese, The Sunway TaihuLight, which performs 93 Trillion operations per second. Unfortunately, an exponential growth in computer capacity has not led to the development of a conscious artificial intelligence.

Alan Turing (1912-1954) was instrumental in the development of computer science and widely considered to be the Father of Artificial Intelligence.

ALAN TURING - FATHER OF AI

Alan Turing's work distinguished between two approaches to AI – the "top down" and "bottom up" categories. Jack Copeland in his article, What is Artificial Intelligence, provides an excellent description of these two basic categories.

> Turing's manifesto of 1948 distinguished two different approaches to AI, which may be termed "top down" and "bottom up". The work described so far in this article belongs to the top-down approach. In top-down AI, cognition is treated as a high-level phenomenon that is independent of the low-level details of the implementing mechanism--a brain in the case of a human being, and one or another design of electronic digital computer in the artificial case. Researchers in bottom-up AI, or *connectionism*, take an opposite approach and simulate networks of artificial neurons that are similar to the neurons in the human brain. They then investigate what aspects of cognition can be recreated in these artificial networks.
>
> The difference between the two approaches may be illustrated by considering the task of building a system to discriminate between W, say, and other letters. A bottom-up approach could involve presenting letters one by one to a neural network that is configured somewhat like a retina, and reinforcing neurons that happen to respond more vigorously to the presence of W than to the presence of any other letter. A top-down approach could involve writing a computer program that checks inputs of letters against a description of W that is couched in terms of the angles and relative lengths of intersecting line segments. Simply put, the currency of the bottom-up approach is *neural activity* and of the top-down approach *descriptions* of relevant features of the task.
>
> The descriptions employed in the top-down approach are stored in the computer's memory as structures of symbols (e.g. lists). In the case of a chess or checkers program, for example, the descriptions involved are of board positions, moves, and so forth. The reliance of top-down AI on symbolically encoded descriptions has earned it the name "symbolic AI". In the 1970s Newell and Simon--vigorous advocates of symbolic AI--summed up the approach in what they called the Physical Symbol System Hypothesis, which says that the processing of structures of symbols by a digital computer is sufficient to produce artificial intelligence, and that, moreover, the processing of structures of symbols by the human brain is the basis of human intelligence. While it remains an open question whether the Physical Symbol System Hypothesis is true or false, recent successes in bottom-up AI have resulted in symbolic AI being to some extent eclipsed by the neural approach, and the Physical Symbol System Hypothesis has fallen out of fashion."[lviii]

Of course, the trillion-dollar question (to match the gargantuan operations per second) is whether all that speed and capacity results in consciousness tantamount to human intelligence. In my view, as will be discussed in the context of Teilhard's Arrow, our most advanced computer systems will never become conscious or approximate human level intellect, let alone a simple insect. The greatest hypothetical obstacle to such development is well described in what is popularly known as the "Chinese Room Objection" formulated by philosopher John Searle.[lix] His premise is that every operation performed by the most sophisticated computer could also be duplicated by a human participant with a simple pencil and pad. He poses the situation that the Chinese unveil a supercomputer which can respond to questions in Chinese. Assume for purposes of this hypothetical that a group of human participants who speak no Chinese occupy a large area inside the computer and simply perform mathematical functions once a proposed question is submitted to the computer and is dropped in the slot. When the Chinese question comes in, the human "clerks" convert the characters to a series of symbols which then correlate with binary numbers. The binary numbers are then correlated with other symbols by means of lists to produce a final numeric product which is then converted into the resulting answer in Chinese which is conveniently pushed through the output slot by the last clerk in the room. An observer on the outside might very well conclude that it was having a human level conversation with a computer (hidden behind a curtain) and thus be unable to distinguish between the computer and human. Such a result would pass the "Turing Test" which was developed by Alan Turing in 1950 to test whether computers could exhibit human level intelligence.[lx] That said, the practical result of the "Chinese Room Objection" is that the computer would be incapable of intelligent thought or discourse as the operations would result from the mindless mechanical acts of non-Chinese speaking clerks lurking inside all those AI bells and whistles.

The bottom-up approach addresses the AI issues from an entirely different perspective than the top-down approach. The concept is to develop the artificial analogues of organic neurons or neural networks which respond to real world phenomena in real time. A simple example would be machines which learn to navigate a room by running into four walls. As the robotic computer encounters the wall or obstruction, the event would be registered as a minute change in the computer program subject to later recall or memory if the same situation recurred. In this way the computer would appear to learn as it encountered stimuli. Many of these bottom-up programs also known as AI connectionism, have become incredibly sophisticated and are used in the field of medical diagnosis and face recognition. However, the same objection confronted by top-down system is equally applicable to the artificial neuron variety. Can these systems beat the Chinese Room Objection and truly be considered intelligent? According to John Searle, the answer is unfortunately, "no", as our human participants could just as easily approximate an artificial neuron or neural network in their mindless assembly work. Yes – the neural network might pass the Turing Test, but intelligent it could not be. The "appearance" of consciousness once again falls victim to the legions of clerks inside the machine.

So where does all this leave us and where does Teilhard's Arrow chime in? Applying the concepts articulated in this work, intelligence is an emergent phenomenon which results from the process of complexification at every point along Teilhard's Arrow. With the advent of organic life, there was an interconnection between the body of the organism (somatic) and development of nervous system and some form of consciousness whether primitive (insect) or sophisticated (human). Certainly, the defining characteristic of human consciousness is the expansion of the brain which was discussed previously in detail. The study of the evolution of man instructs that consciousness and the trappings of culture and science emanated from the organic expansion of the brain and that at a certain point in this continuum of complexity, mankind became aware of itself as a separate entity. In Teilhard's view, this was the birth of thought.

I would posit that although the "bottom-up" modality has not resulted in the development of human level intelligence an awareness, it is moving in the right direction. The key is to develop an artificial system which can exhibit the positive emergent effects of complexification. The early discussion in this work of Artificial Life is instructive but incomplete as a vehicle to foster the development of intelligence at this time. The simple e-organisms can replicate forms and even exhibit RNA qualities at millions of generations.[lxi] However, the emergence of self-replicating e-organisms on one hand and intelligence on the other, is light years apart. That said, it might be theoretically possible to evolve an e-organism into a complex iterations over trillions of generations using our most sophisticated computers.[lxii] Assuming that complexification is an objective principle which manifests in e-evolution, theoretically it should be possible to eventually engender the creation of an intelligent e-organism even at a primitive level, say the neural capacity of an insect or fish. Assuming such an e-organism could be created, it might be possible to e-breed these creatures and expose them to harsh environments to accelerate their evolution over time. In such way it might be possible to ascend the scale of intelligence to the equivalent of primates and even human beings and beyond.[lxiii]

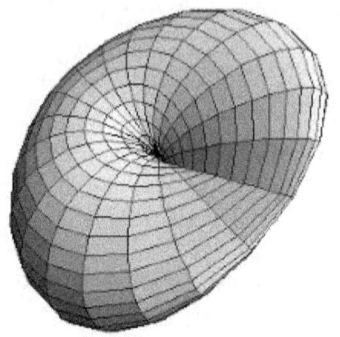

HYPOTHETICAL COMPUTER GENERATED Q MANIFOLD

I predict that the development of quantum computers which run a Game of Life simulation program would give rise to human level intelligence or beyond. The best way to conceptualize this is to view the probability of a computer consciousness arising from the potential configurations of a game of life program. Let us say that the probability of the correct configuration is the same as the occurrence of the cosmological constant – 1 in 10 the negative 122. Even the fastest super computer in the world would need trillions of years to arrive at the winning configuration. However, a true quantum computer process would not be binary but nonlinear as all combinations would be potentially calculated in one instance. Thus, the employment of a quantum computing capability (operating system) to a game of life simulation would create the winning combination within the structure of a computer-generated Q Manifold. Thus, the winning combination to create true AI consciousness as an output of complexification configurations in the simulation would come into existence as the moment the Q Manifold arises.

Therefore, if one predicts that a nervous system would evolve as a result of the interplay of a simulated automata (soma) within the simulated environment, the complexity of the nervous system would increase with complexification. This is the basic tenet of Teilhard's Theory. At a certain point within the realm of probability, even if the number was equal to or greater than the probability of the cosmological constant, a sentient entity would come into existence. Therefore, it is predicted that when quantum computing becomes a reality and is combined with the game of life simulations, a computer general Q Manifold will come into existence wherein human level consciousness analogous or greater will emerge, I would consider the speed of emergence of this computer consciousness with the computer-generated Q Manifold to be analogous to Kurzweil's singularity. At that moment, true intelligence with the sentient power of human consciousness would emerge.

Interestingly, the questions presented in the discussion of the Noosphere might be applicable to our advanced e-creatures. Assuming our e-ecosystem was able to develop a human level intelligence at some astounding height on the generational scale, would not the perpetuation of this intelligence be dependent upon the continued electronic operation of the original computer system which gave birth to our advanced e-organism?[lxiv] Let's say we were conversing with a greater than human level e-consciousness and then accidently pulled out the plug. There is an absolute certainty that the consciousness would vanish along with the power source.[lxv] The cessation of e-life would raise a host of moral questions. Would it not be an act of murder to disconnect a fully conscious e-organism? How would that be different than destroying a human life and would that not constitute an evil act? Assuming the conscious e-organism became aware of its own mortality, would it not attempt to safeguard its power source or store its memory at a myriad of locations. That might protect our e-conscious entity from human destruction and engender a conflict between the human and electronic intellects. Such would recall the ending of the science fiction movie Colossus: The Forbin Project, wherein Colossus and his Russian electronic counterpart protected their power source and forced all of humanity to submit to their will (spoiler alert not required). Sounds plausible to me and reminiscent of the warning articulated by Stephen Hawking.[lxvi]

COLOSSUS TAKES OVER THE WORLD – COLOSSUS: THE FORBIN PROJECT (1970)

Another bottom-up approach which might bear conscious fruit is the early experimentation with combining organic neurons with operating electronic systems.[lxvii] This research is still in its infancy but may offer a way to utilize the existing complexification of organic neurons in tandem with computer programs to generate a hybrid of conscious thought. Referring back to Sheldrake's theory of extrinsic morphic fields or Carl

Jung's collective unconscious, it is possible that the use of neurons within an artificial system could establish a connection to external memory fields. For example, a system which utilizes human neurons might be able to connect to the collective unconscious and gain access to the identical information available to human beings. Interestingly, the knowledge imparted to the e-organism itself would be analogous to the Personal Morphic Field previously discussed. The human handlers, so to speak, could e-raise this e-child and impart experiences and a personal history. That e-organism would then experience not only awareness of self but a defined personal history analogous to any us. However, the interface between the Personal Morphic Field and the Collective Morphic Field would make available to our e-organism all of the information experienced and stored by humankind. Would not such emergent consciousness become self-aware in a truly human sense and also exhibit the capabilities to exceed same?[lxviii]

    In conclusion, it is my opinion that existing top-down systems will never develop human level consciousness. Teilhard's Arrow offers the same solution which led to the complexification and expansion of the human brain. Such process could be employed in a far more sophisticated application of the bottom-up approach utilizing e-life systems or combining organic and artificial neural networks.

CHAPTER SIXTEEN

GENESIS 1.0

Physicist, Stephen Wolfram, provides, perhaps, the most comprehensive analysis of cellular automata in his treatise, *A New Kind of Life*.[lxix] Although Wolfram has not provided a scientific model to explain the complex structures which form and evolve in "Game of Life" programs, he has intuited concepts which illuminate the underlying principles. His basic construct, "Computational Equivalence", offers a paradigm for understanding the existence of platforms both natural and artificial. "The key unifying idea that has allowed me to formulate the Principle of Computational Equivalence is a simple but immensely powerful one; that all processes whether ty are produced by human effort or occur spontaneously in nature can be viewed as computations."[lxx] Wolfram further concludes that the computational systems have common features which make them equivalent or universal. Example of Computational Equivalence include natural phenomena and computer programs which employ cellular automata. A profound extrapolation by Wolfram is that the universal programs can be utilized to produce more complex structures which emerge from the platform. As he states, "And what we have found in this book is that programs are very much the same; some show highly complex behavior, while others show only simple behavior. Traditional intuition might have made one assume that there must be direct correspondence between the complexity of observed behavior and the complexity of underlying rules. But one of the central discoveries of this book is that in fact there is not. For even programs with some very simple rules yield highly complex behavior…".[lxxi]

16.1 OPEN VS CLOSED SYSTEMS

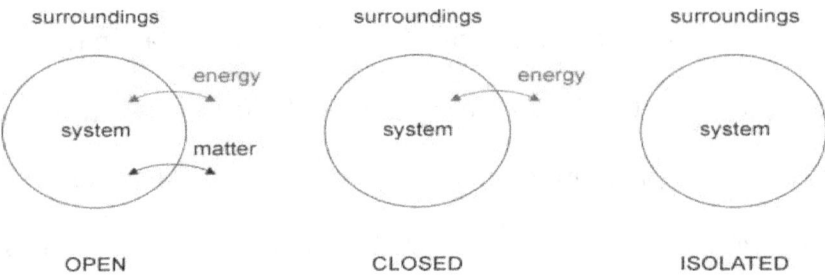

Wolfram does not appear to address what this author terms "open systems" which operate contrary to the computational equivalent systems described in A New Kind of Science. The systems set forth by Wolfram perform discrete functions within a finite timeframe. For example, he considers the computational software of any personal computer a computation which exhibits equivalence with like systems. A system which generate cellular automata would also complete the observed tasks within a discrete period of time. I would characterize these platforms as being "closed systems" as they have defined beginning and end.

That said, Wolfram does not appear to address what this author refers to as an "open system" wherein the platform whether in nature or artificial has a defined beginning but no terminus. Data on these open systems would run computations of equivalence and yet continue to evolve into more complex structures going forward. An example would be a potential cellular automata experiment running indefinitely on a super computer. We have already seen evidence of complex replication structures akin to RNA evolving spontaneously at million of generations. What would trillions of generations bring forth?

An open system by definition would have no end to the evolving complexification of structures. Interestingly, it is this construct which might best explain the underlying nature of complexification observed in all things inanimate and animate – matter and spirit.

For purposes of this work let us assume that the Big Bang itself initiated a platform which exhibits computational equivalence. However, this platform would be an open system. For lack of a better term, let's call this computation "Genesis 1.01". Presumably, the analogous structure observed in "closed systems" would also be observed on this cosmic platform. Like structures would evolve in accordance with the "rules" set up at the beginning of the experiment. The rules of Genesis 1.01 would that pure energy would burst forth at the inception of the computation and condense into elementary particles and matter during the game which originated approximately 14.5 Billion years ago.

The propulsive force which Teilhard refer to as centrification, characterizes the increase in complexity over time. At some point along this platform of cosmic computational equivalence matter makes the centrified leap from inorganic to organic and then to mind and intelligence.

The author's formulation of the Q Manifold might provide a scientifically verifiable explanation as to how complexification in evolution occurs. As previously discussed within the context of Dr. Spetner's research, the statistical chance of even one positive mutation occurring is trillions to one. Accordingly, there would be insufficient time available in a linear progression of natural history over 4.5 Billion years since the origination of the Earth for the plethora of evolutionary changes to occur. However, if one considers all possible combinations of mutations (positive and negative) to be available within the Q Manifold, the winning combination, so to speak, would occur within the total field. In terms of a Game of Life program, it would be as if every mutation configuration played out simultaneously on the quantum computer-generated platform. Accordingly, whether the evolutionary path is from rodent to bat or first vertebrate to man, the winning combination would exist within the Q Manifold. The existence of the Q Manifold, therefore, is the best explanation for evolutionary complexity, not natural selection which would require trillions of years to play out to achieve the successful configuration of multiple random mutations. A linear progression of 4.5 Billion years of Earth history is insufficient to arrive at the positive evolutionary changes as previously discussed.

To be more specific, the Genesis 1.01 program would allow for Game of Life complexification of life on Earth to exist within an "open system". As the combination for complexification and successful random mutations would exist at a preset probability, it would still occur within the Game of Life simulation even if the odds were 10 to the negative 121. *Critically, the probability of a successful mutation configuration is never zero and that is the salient conclusion.* With sufficient quantum calculative power, the combination would be arrived at which would never arise within our current binary operating systems.[lxxii]

Finally, if we consider the manifold as an asymmetrical object of which space is the outer shell or surface, the winning combination of atoms would exist at a defined point in the manifold in the same way the probability of the Cosmological Constant would arise at a point between the countervailing geometric effect of expansion of the manifold (perceived as expansion of the universe) versus the curvature of space in the presence of mass (gravity).

16.2 GRAVITY WELLS AS DRIVING FORCE OF COMPLEXIFICATION

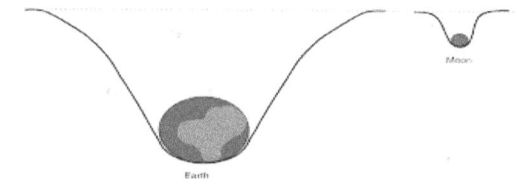

GRAVITY WELL

Einstein critically changed our understanding of gravity as originally postulated by Isaac Newton which relied on the propensity of objects to exert attractive force. Einstein changed all of that theorizing that the geometry of space was responsible for human observation of gravity. Mass causes space to curve in proportion to the density of the matter presented. The greater the mass – the greater the curvature of space and also time.

The evolution of starts and planets from primordial gas is illustrative of the effect of mass on space. The non-Einsteinian theory suggests that particles are attracted to each other resulting in the aggregation of mass from minuscule amounts to massive objects such as stars and galaxies.

However, the true cause of such aggregation of mass results from the curvature of space and the accumulation of mass within the confined area of a gravity well. As more mass is deposited so does the curvature of space increase allowing for the further increase in mass accumulation. Once again, we have described a positive feedback loop.

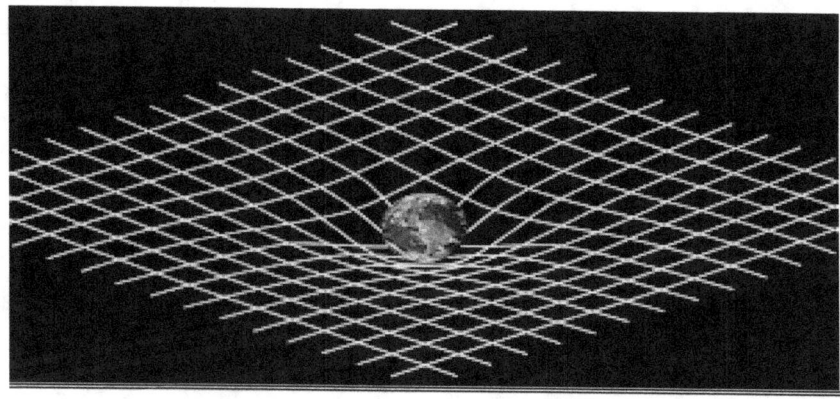

CURVATURE OF SPACETIME

I would suggest that the resultant gravity wells are breeding grounds not just for the complexification of matter but the emergence of life itself in the universe. These gravity wells would spawn the emergence of stars and planets and ultimately life as complex structures evolve within an open system of computational equivalence. I would posit that the centrification which propels complexification forward in a Teilhardian context is a direct result of the curvature of spacetime. As such, human intelligence are emergent byproducts of the rules being played out under Genesis 1.01.

If this observation is correct, one would expect to find the origination of a noosphere and omega in any cosmic computational equivalent system. Carrying this thought experiment forward it is entirely possible that a myriad of noospheres and omega exist within the confines of the gravity wells scattered across the universe.

An intriguing question would be to postulate what level of intelligence might exist in the ultimate curvature of space – a black hole. The likely outcome, however, would be that complex structures cannot survive the brutal compressive force of a black hole.

Intuition inform that the universe is replete with the most complex and sentient consciousness existing within the gravitational belly of a gravity well.

CHAPTER SEVENTEEN

THE ANATOMY OF EVIL

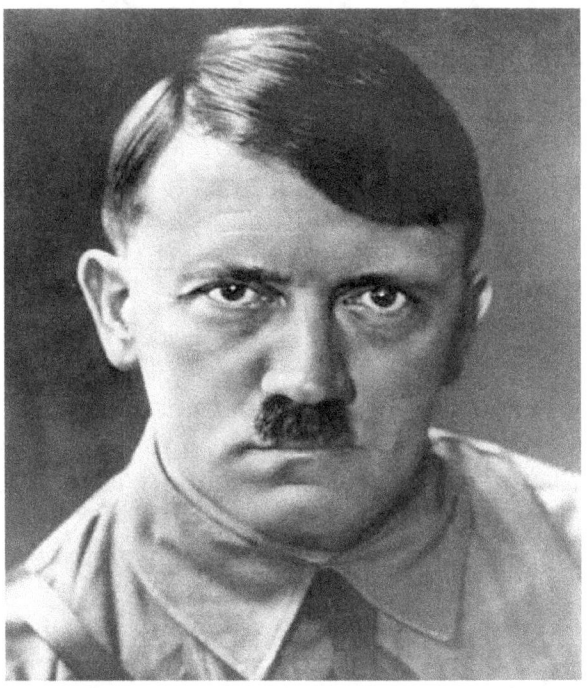

Teilhard refers to the presence of evil or suffering as the manifestation of entropy, i.e., the disintegration of ordered states over time. Suffering and death are viewed as the natural progression. Matter, whether exhibiting characteristics inorganic or organic, eventually submits to the "evil" of entropy. Teilhard states, "To begin with, we find physical lack of arrangement or derangement on the material level; then suffering which cuts into the sentient flesh; then still on a higher level, wickedness and the torture of spirit as it analyzes itself and makes choices. Statistically, at every degree of evolution we find evil always and everywhere, forming and reforming implacably in us and around us."[lxxiii]

The very concept of the Noosphere and the final progression of Omega provide for the emergence of a supernatural presence as a result of the complexification of matter. It is my speculation that some emergent centrification of pain, death, and the human concept of "evil" exists within the structure or landscape of the Noosphere as postulated by Teilhard in the same manner as certain sensation such as touch or smell reside in areas of the brain. Thus, would evil or violent tendencies exist and be present within the confines of preset Jungian archetypes. Evil, therefore, is very real as a force outside of human comprehension and *not* simply a metaphor for a physical process, to wit, death or entropy.

"Thought" according to Teilhard constitutes matter's ability to reflect on itself. Further complexification provides for the emergence of a collective thinking mass of all mankind and the ultimate centrification of same as reflected in the Omega Point itself. Whether we refer to this ultimate iteration of matter as Omega, the Father, Christ, or Yahweh, it inhabits a state which must be viewed as supernatural or super phenomenal from our inferior perspective as homo sapiens. The larger question to ponder is whether Omega as an omnipotent force or entity is *benign*? Given the equality and duality between the forces of centrification and entropy, why would the attributes of the former be the sole province of existence at the Omega Point? At all inferior layers along Teilhard's arrow, inorganic and organic life exhibit the duality of complexification within the backdrop of disintegration, death, and entropy.

My key observation is not simply the presence of the duality at the organic and inorganic level but that this "duality" is also manifested in higher emergent states as one ascends Teilhard's Arrow. For example, it is axiomatic that all atomic structures eventually decay. Uranium 234 has a half-life of 245,500 years. Iron, with a more stable atomic structure, decays over a period of 2.6 million years ultimately devolving into Cobalt-60. The disintegration or decay of atomic structures reflects the entropic process at an elementary level along Teilhard's arrow. The destruction of form can be viewed as the phenomenon of death, but at an inferior level.

The emergence of organic life as a byproduct of complexification also led to the objectification of duality but at a higher level of manifestation. Cells have varying lifespans based on function. For example, a skin cell has a lifespan of three weeks whereas a red blood cells survives for four months. The destruction of this organic form can similarly be viewed as the phenomenon of death, but at a higher nested level than its atomic sub-components. That said, the exposure of unicellular structures such as an amoeba to noxious substances will result in what appears to be a writhing sensation.

Ascend to a higher level of complexification along the continuum of organic evolution and the manifestation of duality becomes more pronounced. Vertebrates and some invertebrates such as squids exhibit pronounced evidence of pain which precedes the cessation of life whether by natural consequences or the acts of predators.

With the emergence of thought as a byproduct of consciousness, the duality of life and death becomes a core feature of our human experience. Human beings, of course, die like any other biological structure. The huge difference, however, is that we all "know" that we will die at some point. Death, therefore, finds a higher level of expression within the emergent state of human consciousness. In the same fashion, the experience of pain is also amplified by our conscious state. We not only exhibit the presence of pain, but we are aware that we are experiencing this sensation which amplifies its effect. The emotions we experience as human beings add a nuanced value to pain and death. We dread pain and do our best to avoid it. Most of us fear death and we do our best to prolong life notwithstanding the pain experienced in the final stages of disease.

As stated previously, we cannot predict a successor emergent state from our inferior position along Teilhard's Arrow. That said, every emergent state demonstrates the existence of the dual forces of complexity and entropy. Therefore, one would expect that this duality would be manifested at the level of the Noosphere. At the higher emergent level of the "thinking mass", complexification and entropy would also be existent. One can only speculate as to how the Noosphere would experience pain and death. Its higher level of consciousness as a sentient force beyond the level of individual minds would demonstrate a consciousness superior to individual humans. However, since the thinking mass would accrete or abate based on the finite lifespan of humans, the duality of life and death would be expressed somehow at a level beyond our inferior position on Teilhard's arrow. Beyond the presence of life and death, how would pain be manifested within the Noosphere and would the presence of "evil" be present in some capacity?

With the advent of human consciousness, the phenomena of pain and death are exhibited and experienced at a higher level of complexification. Human beings can consciously and with volition cause pain and death to other humans. Although mammals will hunt and kill prey for food, murder appears to be limited to human beings and possibly the higher primates such as chimpanzees and apes.[lxxiv]

NUCLEAR BOMBING OF HIROSHIMA – AUGUST 6, 1945

Within the human mind, the ability to inflict pain is amplified by intelligence and evidenced in the evolution of man. The murderous activity of mankind goes back thousands of years and has even been identified among the population of Neanderthals. In the past 100 years, humans have developed ever more sophisticated methods of causing pain and death. With the advent of nuclear weapons, we now possess the ability to cause untold pain and even the extinction of our species. Incidents of "evil" behavior are, unfortunately, well documented. Hitler exterminated millions of Jews and other ethnic groups. Evil as a demonstrable force, was certainly evidenced in the thought process of Hitler and others like him. In the same vein, acts of charity and self-sacrifice are equally prevalent among the heroes in our daily life and luminaries such as Mother Theresa.

16.1 WOTAN AND THE GERMAN ARCHETYPE OF EVIL

HIERONYMUS BOSCH PAINTING

Jung believed that a repressed capacity for evil existed within the archetype of Wotan, the ancient German God of War. The manifestation of this evil archetype may have been responsible to the rise of Nazism and the resulting death of millions of Jews and other ethnic groups during the reign of Hitler. Jung believed that this archetype and others had a huge effect on human populations. "...they exist and function and are born anew with every generation. They have an enormous influence on individual as well as collective life and despite their familiarity they are curiously non-human. This latter characteristic is the reason why they are called Gods and Demons in the past and why they are understood in our 'scientific' age as the psychical manifestations of the instincts, in as much as they represent habitual and universally occurring attitudes and thought forms. They are the basic forms, but not the manifest, personified or otherwise concretized images. They have a high degree of autonomy, which does not disappear, when the manifest images change."[lxxv]

From a Teilhardian perspective, I would posit that this Wotan Evil Archetype is akin to a computer file within the Noosphere database (collective unconscious). It acquires information from the learned behavior of violent humans over the millennia in the same way McDougall and Crew's mice accessed the collective memory of the Noosphere. As the file grows in size and magnitude over time, the behavior is immediately learned by successive generations and ultimately manifests and objectifies as human behavior, to wit, violent acts. It exhibits the features of accelerating returns or positive feedback as postulated by Kurzweil's singularity. As more violent behavior is stored, there is a greater propensity for violence in successive generations. Violence thus begets violence. I would further suggest that humans have the ability to contemplate a wide spectrum of possible behavior or acts. It may be that the very process of visualizing or imagining violent behavior opens the Individual Morphic Field to the file download of violent stored knowledge in the Wotan Evil Archetype or applicable construct. In the same way an artist or physicist may input (intuit) memories of universal concepts by accessing an archetype or area of the Noosphere, evil thinkers access their own malignant archetype and stored information. Consequently, the human volitional element or choice might trigger the evil act or consequence.[lxxvi] Teilhard seems to suggest this when he states, "to begin with we find physical lack-of-arrangement or derangement on the material level; then suffering, which cuts into the sentient flesh; then, on still a higher level, wickedness and torture of spirit as it analyses itself makes choices."[lxxvii] As Intelligent Process presupposes the co-existence of nested levels of increasing complexity which contemporaneously spans the quantum to the inorganic to the organic to the conscious to the Noosphere to Omega, death, pain, and evil are built into the very fabric of matter. Therefore, evil exists at every level because entropy and synthesis, driven by the force of complexity, is the very spine of the phenomena of Intelligent Process. With the advent of mankind and thought, the presence of death and pain take on a larger meaning. As above stated, we consciously perceive the existence of pain and death in a manner which is superior to life at lower levels of complexity. Given that mankind can act with intent to cause pain and death, our very existence as conscious entities allows for the concept of "evil" to emerge. Therefore, at our level of emergence and complexity, the phenomena of evil is actualized since it is by nature a byproduct of human thought and reflection. Evil, therefore, is literally built into the fabric of our existence at our level of emergence along the continuum of Teilhard's arrow.

16.2 POST-<u>NOOSPHERIC SPECULATION</u>

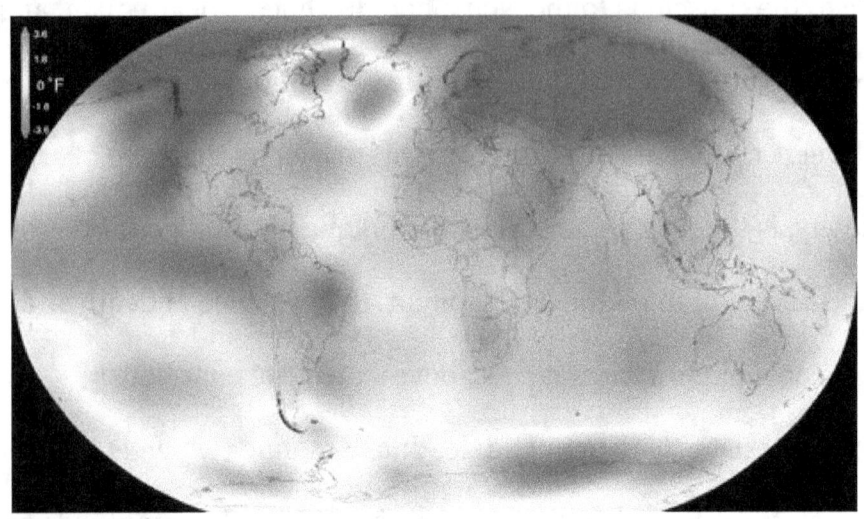

<u>FIRE OR ICE – WHICH WINS OUT?</u>

Once again, the most fascinating speculation is the characterization of God as the level of emergence beyond the Noosphere, the Omega Point. As hypothesized, this synthesis might manifest intelligence which is separate and apart from its biological underpinnings and thus exist independently of the life or death of humankind. This underlying assumption which is, of course, pure speculation, is critical to the final conclusion as to the existence of evil. Since the Omega or Omega Point would exist independently of its biological origins, the very process of entropy would ultimately be defeated, and synthesis would win out. The very process of death would cease to exist at Omega along with its accompanying elements of death and the human conceptualization of evil. Stated more simplistically, God would trump Evil and a presence akin to the devil or some dark manifestation would logically not exist at the Omega Point.

All this speculation brings us back to a question posed by philosopher, David Hume as to the culpability of God for the very existence of evil in the world. The answer is that volition has nothing to do with it. Systemically, evil in some emergent form would have to exist within the Noosphere. The very existence of God or Omega presupposes the prior nested levels which always includes some analogue of pain, death, and certainly the awareness of evil as per conscious thought. In the same manner that water cannot negate the existence of its component parts, Omega cannot defy the predecessor forms which engendered its very existence. Without the existence of evil, therefore, Omega would not be possible. Evil exists because God and Omega contemporaneously exist.

God, therefore, is not impotent or malevolent, but simply exists in some sentient form or structure beyond our level of comprehension along Teilhard's arrow.

Conversely, assuming that Omega is not independent of its biological origin at lower nested levels of complexity, the very notion of evil would find some analogue within the hyper-conscious state of Omega. God, therefore, utilizing this paradigm, would exist as a hyper conscious force beyond the mentality of the thinking mass of the Noosphere. That said, it would also manifest a hyper centralized manifestation of evil as well as synthesis. The concepts of evil and good would thus exist contemporaneously at the Omega Point in this alternative set of assumptions. The outcome of Hume's inquiry, however, would be the same. Omega could not defy itself and there would be no volitional mechanism to prevent or allow for the existence of evil.

All of this brings us back to the ultimate questions as to why evil exists in human society? It exists because it is part of the nested fabric of who we are at our level of emergence on Teilhard's arrow. The existence of evil is also mandated because a higher level of complexity which also embodies evil in some fashion must exist contemporaneously within the Noosphere as discussed in detail. Hitler as a historical occurrence is tragically a byproduct of who we are and will likely recur. In the same fashion we can expect that there will be continual acts of kindness which represent the embodiment of synthesis and a greater good. Mankind cannot rid the world of evil because it is built into the fabric of who we are.

Author, Ron Rosenbaum in his book, Explaining Hitler, the Search for the Origins of His Evil,[lxxviii] considers the philosophical discussion of the very foundation of evil. Citing the work of Berel Lang, Rosenbaum characterizes the nature of evil thus: "To do evil despite knowing its evil is one thing… But to do it because it is evil is quite another, isn't it?"[lxxix] Therefore, pure evil dictates by its core a "consciousness of the evil nature of an act."[lxxx]

In my view, evil exists at the very heart of human consciousness and is the thought manifestation of Teilhard's concept of disorder. The centrification of this principle at the level of consciousness results in the performance of evil acts. Recall that Teilhard concludes that "Statistically, at every degree of evolution, we find evil always there…" [lxxxi] This observation is important because it disputes the liberal revisionist view that evil acts have some explicable source or causation based on scientific research. This approach is embraced by author, Simon Baron-Cohen in his book, The Science of Evil – on Empathy and the Origins of Cruelty.[lxxxii] In attempting to explain the horrific acts which led to the Holocaust, Baron-Cohen disputes that evil is an independent force. "Why did the murderer kill an innocent child? Because he was evil. Why did the terrorist become a suicide bomber? Because she was evil. But when we hold in the concept of evil to examine it, it is no explanation at all. As a scientist, I want to understand what causes people to treat others as if they were mere objects."[lxxxiii] Baron-Cohen then digresses to a proposed explanation based on what he calls "an erosion of empathy" and ability to dehumanize and treat people as objects.[lxxxiv] This so-called scientific approach seeks to rationalize the existence of evil as a causal effect of some extrinsic factor such as psychosis, sociopathic behavior, or genetics. These explanations are not plausible, in my view, as evil statistically emerges in diverse populations regardless of genetic heritage or upbringing. Children can be raised in identical environments with diametrically opposed outcomes. Another way to look at the existence of evil is to postulate that evil acts have occurred since the beginning of history and have are part of the systemic structure of the Noosphere. As evil exists within the Noosphere as part of the collective consciousness it must be accessed by individuals at a lower level of emergence. The existence of one necessitates the existence of the other. Tragically, the increased occurrence of evil acts might increase the strength or attraction of the learned behavior in the same way that certain knowledge is proceeding rapidly to a singularity. The increased frequency of evil in the world may be rocketing literally and figuratively towards its own evil singularity which sound as foreboding as the End of Days.

16.3 DISTINCTION BETWEEN OUTWARD STRUCTURE OF NOOSPHERE, THINKING SPHERE OF NOOSPHERE, AND OMEGA

The anatomy of evil may be better understood by clarifying the distinctions between the structure of the Noosphere, the Thinking Mass of the Noosphere, and Omega itself.

a. Outward Structure of Noosphere.

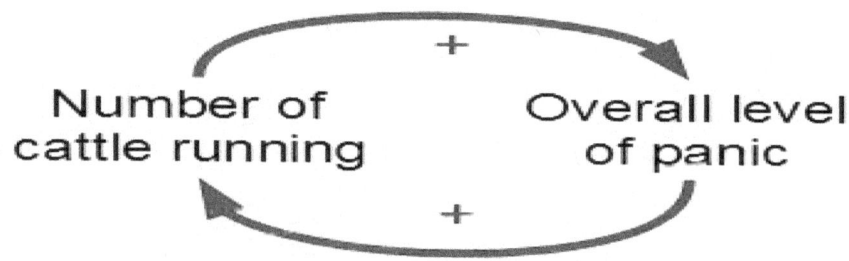

POSITIVE FEEDBACK LOOP – "TEXAS STYLE"

As previously discussed, violent and malevolent knowledge may be stored in discrete areas of the Noosphere akin to regions of the brain. Whether characterized as an archetype or sub-file within a computer analogous to the Collective Unconscious, a history of human violence would accrete over time and gain dominance. The Kurzweil positive feedback loop would apply and the power of dominance of this storage of evil information would become increasingly accessible to the human population akin to the research of McDougall and Crew's laboratory mice. The frequency of evil acts would thus increase overtime in addition to its presence within distinct populations. An underlying feature of this retention and access to malevolent knowledge would be the transference of information from the Collective Morphic Field to the Individual Morphic field.

b. Evil and the Thinking Mass of the Noosphere.

VISION OF EVIL BY HIERONYMUS BOSCH

Evil within the Thinking Mass of the Noosphere would be an emergent feature of the actual storage of information within a discrete region of the Noosphere categorized by a discrete malevolent archetype at the emergent level of the Noosphere's Thinking Mass. I refer to such emergent feature as "Noospheric Consciousness" which would be analogous to human consciousness which is qualitatively distinct from the storage of information with specific regions of the brain. Noospheric Consciousness (as an emergent feature beyond the level of human consciousness) would be aware of its sub files of information within the outward structure of the Noosphere and incorporate same as the whole greater than the sum of its parts. Noospheric Consciousness would be at a superior level to the reflection of thought at the human level. *As matter became aware of itself at the threshold of thought according to Teilhard, so would thought become aware of itself at the higher level of emergence at the Noosphere* (according to Moster). As above referenced, Teilhard embraces such hyper consciousness when he states, "to begin with, we find physical lack of arrangement or derangement on the material level; then suffering which cuts into the sentient flesh; then still on a higher level, wickedness and the torture of spirit as it analyzes itself and makes choices."[lxxxv]

The Noospheric Consciousness might also set the stage for what we humans refer as the battle between good and evil – angels and demons. A Thinking Mass aware of its sub files or archetypes of good and evil could choose to pursue either of these alternatives. At this higher level of emergence, in the way hydrogen and be combines with oxygen to form water, a Noospheric Consciousness could select evil ends and guide the transference of malevolent information from evil archetypes through intent at this higher level of emergence. Conceivably, this dynamic would allow for the existence and propagation of pure evil as its own sentient force akin to the Christian Bible's description of the Devil. In this wild conjecture and thought experiment, the sentient evil force would choose human subject to manifest evil, i.e., Hitler, mass murderers, etc. Further study of the potential existence of demonic possession and exorcism might offer scientific proof of this hypothesis.

In the same fashion, pure "good" would also reside in its own archetype and be present in the Noospheric Consciousness as a sentient angelic force. Such angelic intelligence would be as equipped as its demonic counterpart to impart knowledge to human subjects.

The juxtaposition of Noospheric Consciousness in the realm of good and evil would play out on the human stage with the texture and content of the bible itself which would not be viewed as a mere metaphor but a literal accounting.

c. **Evil & Omega**

GOOD & EVIL – PORTRAIT BY DEMETRIO AGUILAR

At the emergent level of Omega, thought would turn on itself "thrice". Teilhard, of course, posited that human consciousness represents the first-time matter reflected on itself ushering in the threshold of thought. This author posits that the Noosphere's Thinking Mass would represent a Noospheric Consciousness or the emergent nature of thought becoming aware of itself as an independent entity, to wit, all the knowledge of collective consciousness or the Thinking Mass becomes aware of itself and conscious. The consciousness that succeeds the Noospheric Consciousness would be Omega, a state that cannot even be fathomed from our inferior level.

Assuming that the very existence of Omega is biologically predicated on the existence of the lower nested organic levels, Omega would derive its very existence from the structure of matter at all prior levels, to wit, entropy and synthesis. Thus, evil would live on in some fashion in Omega itself as entropy would find some expression at the very summit.

Conversely, if Omega is distinct from predecessor biological levels, it would exist as a sentient *sentient* force beyond the limits of prior biology, entropy and death. In this scenario, death would be vanquished and good would prevail in Omega. In biblical terms, the Devil would lose the battle as prophesized in Revelation.

Conceivably, this Noospheric Intelligence would manifest as a sentient force of evil and seek to manipulate human at the level of the collective unconsciousness and Individual Morphic Field.

d. Scientific Proof of Noospheric Conscious Intervention and the Advent of Evil.

EXORCISM IN POPULAR CULTURE – EXORCIST STAIRS FROM THE FEATURE FILM LOCATED AT GEORGETOWN UNIVERSITY

As previously discussed, Noospheric Consciousness would constitute a sentient force capable of intervening directly in the physical and mental state of matter at inferior levels along Teilhard's Arrow. At the level of human consciousness, we have the ability to manipulate matter at inferior levels through the process of thought and reflection. Human technology is a perfect example whereby we can create new forms of matter, antimatter, and even artificial intelligence. The source of such intervention at junior levels of matter can be benevolent (say, invention of vaccines and genetic cures) or malignant (biowarfare and nuclear weapons). The key element is that conscious reflection allows for volition – the ability to make a choice as to an outcome.

The second factor is that the outcome is either beneficial or malignant and embodies the underlying principles of complexification, entropy and synthesis. Human consciousness by its very nature is engaged in a battle between good and evil ends. Access of information stored in the outward structure of the Noosphere likely enhances this process of transference of information from the Noospheric sub file (archetype) to the Individual Morphic Field. This would be subject to the positive feedback loop discussed previously based on the research of McDougall, Crew, and Kurzweil.

Noospheric Conscious Intervention would take conscious "intervention" to an emergent and higher level. Noospheric Consciousness would exhibit behavior akin to human consciousness as it would be aware of itself as a sentient force and capable of choosing its objectives and ends. In the same manner, humans can manipulate matter to achieve beneficial or malevolent ends, so would the Noospheric Consciousness. The difference is that the Noospheric Consciousness would be able to manipulate human beings directly

e. **<u>Objective Evidence of Noospheric Consciousness and Evil Intervention.</u>**

This subject seemingly requires a digression into the speculative world of the occult or worse. Towards that end, I will offer up a highly respected source of information on the topic of demonic possession which, by definition, embodies the objective intervention and manifestation of evil on earth. Robert Gallagher, MD is a board-certified psychiatrist and professor of psychiatry at New York Medical College. He received his medical training at Yale University and Columbia. Over a period of 25 years he studied several hundred incidents of alleged demonic activity. Initially approaching the subject with great skepticism, the cases seemed to have a scientific explanation as relating to episodes of severe psychosis and mental illness. However, Dr. Gallagher witnessed actual occurrence which he concluded were proof of demonic possession. There was no other scientific explanation. As he stated:

"But I believe I've seen the real thing. Assaults upon individuals are classified as "demonic possessions: or as the slightly more common but less intense attacks usually called oppressions". A possessed individual may suddenly, in a type of trance, voice statements of astonishing venom and contempt for religion, while understanding and speaking foreign languages previously unknown to them. The subject might also exhibit enormous strength or even the extraordinary rare phenomenon of levitation. (I have not witnessed it myself, but a half dozen people I work with vow that they've seen it in the course of exorcisms.) He or she might demonstrate "hidden knowledge" of all sorts of things – like how a stranger's loved ones died, what secret sins she has committed, even where people are at a given moment. These are skills that cannot be explained except by special psychic or preternatural ability.[lxxxvi] I have personally encountered these rationally inexplicable features, along with other paranormal phenomena. My vantage is unusual: As a consulting doctor, I think I have seen more cases of possession than any other physician in the world."[lxxxvii]

16.4 DOCUMENTED INCIDENTS OF EVIL ACTS

CHARLES WHITMAN

Consider the mass murders perpetrated by Charles Whitman who shot and killed 17 people from the observation tower at the University of Texas on August 1, 1966. Analysis of Mr. Whitman's early life provided no evidence or indication of his subsequent evil behavior. He was described as polite and a person who seldom lost his temper. At age 11, he joined the Boy Scouts and ascended to the ranks of Eagle Scout. He was an accomplished pianist and had a paper route as a boy. Indeed, Mr. Whitman better personifies the traits of Beaver Cleaver of "Leave it to Beaver" fame, then a mass murderer. A reductionist or scientific approach will never explain why Mr. Whitman became a mass murderer. The author suggests three possible explanations for this seemingly inexplicable evil behavior.

First, Mr. Whitman may have acted as he did simply because evil is statistically present in any given population as postulated by Teilhard. It is simply a question of probability and outcome. Evil as an independent phenomena manifests in human consciousness and population on a statistical basis in the same manner as a virus or disease is physically present. Second, Mr. Whitman, through his thought process or some induction within his Individual Morphic Field, may have accessed the stored knowledge of the Evil Archetype. This process would rely on an active transference of information from the Noospheric sub file (archetype) to Mr. Whitman's personal consciousness (his Individual Morphic Field). And third, Mr. Whitman may have been manipulated by the superior sentient evil forces at a higher emergent level through the above process of Noospheric Conscious Intervention.

OSAMA BIN LADEN

Yet another example is that of Osama bin Laden who masterminded the 9-11 terrorist attacks on the World Trade Center resulting in the death of 2,996 innocent people. Such act was pure evil and yet was not indicated in Osama bin Laden's childhood which was rich and privileged. He was born into a billionaire family in Riyadh, Saudi Arabia. Notwithstanding the advantages of wealth and success, Bin Laden was responsible for the death of thousands of innocents in a pure act of evil. No other member of the Bin Laden family which continues to own and operate the wildly successful Bin Laden Group conglomerate, has ever committed such evil acts. It was relegated to a single member of the family. My explanation is identical. Evil is statistically present and simply emerged within the psyche of Osama bin Laden. It may be that his embracement of violent extremism as a concept may have opened the floodgates of the evil archetype which ultimately objectified as the 9-11 terrorist attacks on the United States. He may also have been utilized as a vehicle to accomplish evil through the process of Noospheric Conscious Intervention.

STEPHEN PADDOCK

Stephen Paddock fired semi-automatic rifles with bum-stock attachments from his room on the 32nd floor of the Mandalay Bay into a crowd of 22,000 concert goers on October 1, 2017. The incident resulted in 58 fatalities and became the deadliest mass murder in modern American history.

Paddock had no prior violent history and only interaction with police enforcement was a minor traffic citation years before the shooting. He previously worked for the Post Office, an accountant, and real estate investor. Although an avid gambler, he was a successful entrepreneur and worth millions at the time of his death. He planned the attack with precision having checked into the hotel six days before the shooting and amassed 23 firearms to carry out the murderous act.

Recent information reveals that Mr. Paddock began amassing firearms decades before the Las Vegas Shooting. This conduct may have opened access to the evil archetype which objectified as the mass murders. The remaining two possible explanations for the occurrence of evil similarly apply.

The list of evil perpetrators is unending, and a scientific or psychological cause can rarely be proven. Teilhard is correct as to the existence of evil and it is simply woven into the fabric of matter and a volitional display within the context of mankind.

CHAPTER EIGHTEEN

INSIDE HEAVEN'S GATE

DANTE'S PARADISO

To my faithful readers who have taken the Teilhard plunge until the point, an important question remains unanswered: Does Teilhard's Arrow allow for the existence of Heaven in the biblical sense of the word? Of course, the world religions have taken varying interpretations as to the nature of Heaven and whether it represents an actual spiritual place that we can inhabit after death.

For Christians, the traditional view is that Heaven is an actual location or Kingdom where Jesus and the Father exist along with a bevy of Angels. This view is certainly captured in the intricate artwork of the Dark Ages and Renaissance. John 3:16 sets forth the proposition of eternal life after death in black and white: "For God so loved the world that he gave his one and only Son, that whoever believes in him shall not perish but have eternal life."

Whether these words are taken literally or figuratively depends on one's particular view of Christianity. For example, among my many Southern Baptist friends in West Texas, there is no question that Heaven is an actual location and that those who believe in Jesus as the savior are guaranteed eternal life. That said, numerous Christians take a more attenuated view that the Christian bible is not to be read literally but as a moral guide for living. Teilhard, of course, was a Jesuit and espoused an ardent belief in Christ as a supernatural entity at the apex of creation – Omega.

Buddhism and other eastern religions view Heaven as a temporary abode wherein souls traverse the path from death to rebirth. The concept of heaven, therefore, is cyclical in nature and allows for the birth – death - rebirth in a quest to achieve enlightenment.

In the Jewish faith, it is difficult to ascertain whether Heaven is actual or metaphorical. The Torah provides little insight as to the underlying nature of heaven. Having been born and raised in the Jewish faith, I do not claim to be an expert on whether life after death is a viable concept in Judaism. However, I will report that many of my family members who were facing the consequences of their last breath did not aspire to anything permanent beyond a Jewish burial or cremation. This created considerable anxiety as you can imagine.

From the perspective of this writing and scientific belief in complexity and emergent states, I believe that heaven is an emergent state which is occupied by all of us contemporaneously with our existence on earth. Going back to my use of hydrogen and oxygen atoms which exist as distinct particles contemporaneously with the water molecule, I believe this same recurring pattern of complexity and emergent states at different levels also applies to all of us. In a Teilhardian sense, our individual thoughts are contemporaneously part of the next emergent state known as the Noosphere. Therefore, we exist seemingly as individual entities but at the same time are components of a larger and greater entity which also exists contemporaneously as the Noosphere. Similarly, the Noosphere would have its own contemporaneous centrification which would result in the final iteration – Omega. If Omega exists as I write (which I believe), it constitutes the Final Emergent State wherein all prior sub-states or prior emergent states exist with human individual thought being at a lower level.

Accordingly, I would posit that all of us exist on Earth and in Heaven at the same time and that the pathway from life to death to afterlife in Heaven is not a linear progression but simply different emergent levels of complexity rising from simple matter to organic to primate to human to Noospheric to Omega. Everything exists contemporaneously.

But all of this seems to beg the question of concern to all of us who face an actuarial termination. We are all going to die some day at some defined time. Will our mind or thought process survive as an independent unit after we die?

My sense is that the answer is "yes" from my perspective on Teilhard's Arrow. The very existence of the Noosphere seems to indicate that our memories are preserved within a larger sphere of information which is stored and accessed over time. The reports of reincarnation events discussed in this work which have been scientifically validated, provide evidence that our individual memory continues on beyond death and can be accessed by our descendants and others. Whether this means that our memory is simply preserved as data points or an integrated consciousness that still retains a sense of "self" and "identity" after death is pure speculation.[lxxxviii] My sense is that we would move from the emergent state of our existence on earth as conscious beings to the next emergent state of which we would be a part but would manifest as the whole greater than the sum of its parts – the Noosphere. We might wake up after our last breath as a conscious integrated part of a greater consciousness – Noospheric Consciousness. As this is the next emergent state, we lack the capacity to even fathom what this state would be like. The point is that the Noosphere and Omega cannot exist without the prior existence of less complex (junior) emergent states. In that complexity and successive emergent states characterize all matter and organic forms, the continued existence of prior states must be preserved to allow for the greater emergent state to exist at all. Therefore, the preservation of our memory seems to be assured if one accepts the perspective suggested by Teilhard's Arrow.

CHAPTER NINETEEN

POSTSCRIPT

"SOMETIMES WHEN I'M FACED WITH AN ATHEIST, I AM TEMPTED TO INVITE HIM TO THE GREATEST GOURMET DINNER ONE COULD EVER SERVE AND WHEN WE HAVE FINISHED EATING THAT MAGNIFICENT MEAL, TO ASK HIM IF HE BELIEVES THERE IS A COOK"

RONALD REAGAN

Ronald Reagan had a great way of simplifying concepts. Here's another – I am a terrible housekeeper no matter how hard I try. The sink eventually overflows with dishes, laundry pours out of the closets, and garbage takeout day is on its fourth extension. Refusing to succumb to this onslaught of disorganization, I exert hours of appropriate effort to wash, shine, and throw out and return the inanimate to its prior and customary position. For a day or two, I admire this energy expenditure with some comfort only to confront the same decline. Moral of the story – It takes concerted intent, energy, and determination to maintain order (synthesis) in the wake of disintegration (entropy).

Travel to a location far from the city lights and gaze up at the brilliant stellar display. Only the most limited minds among us could conclude there is no intelligence there. And how about the *precision* of mathematical laws, the *exact* velocity of the speed of light and *not one nanosecond greater*, and the infinite hierarchy of "Order" from the infinitesimal to the gargantuan? Some sentient force must be tending to all of that cosmic housekeeping.

The universe abounds with intelligence. There can be no other explanation.

And so, I have attempted with this interplay of logic and the varied disciplines of Cosmology, Physics, Biology, and Computer Science to prove up the logical existence of God and (ominously) His nemesis. At the very least, I hope that this work has advanced the debate and that the reader can discern, however, tenuous, a bridge between the theological and

scientific, the Material and the Divine. Certainly, Teilhard could, and most definitely, Mr. Reagan accorded proper attribution for his gourmet meal.

END NOTES

[i] Thomas Mulvihill King, S.J. (born 5.9.29 – died 6.23.09). Father King was a beloved and brilliant theology professor at Georgetown University who was an inspiration to generations of students including this author.

[ii] Thomas Mulvihill King, S.J. wrote several books on the life and writing of Teilhard de Chardin including *Teilhard's Mysticism of Knowing* (1981); *Teilhard and the Unity of knowledge* (1983); *Teilhard de Chardin* (1988); *Letters of Teilhard de Chardin to Lucille Swan* (1993); *Teilhard's Mass* (2005).

[iii] *The Phenomenon of Man* (Pierre Teilhard de Chardin; London: Collins; 1966; First French ed.; 1955).

[iv] *Ibid.* at p.254.

[v] Dawkins, Richard. (2006) *The Blind Watchmaker: Why the Evidence of Evolution Reveals a Universe Without Design* New York: Norton.

[vi] For an excellent discussion of this topic see, *Emergence, the Connected Lives of Ants, Brains, Cities, and Software;* Steven Berlin Johnson, 2001, Simon & Schuster.

[vii] For further reading see, *Creation, Life and How to Make It*, Steve Grand, 2001, Harvard University Press; *Artificial Life, A Report from the Frontier where Computers Meet Biology*; Steven Levy, 1993, Vintage.

[viii] As recounted by Baretti, Giuseppe (1757), *The Italian Library Containing an Account of the Lives and Works of the Most Valuable Authors of Italy*, London, p.52.

[ix] *Not by Chance* (Dr. Lee Spetner; New York, The Judaica Press, 1997, 1998). This work subject's classic evolutionary theory to the rigors of science and demonstrates the inadequacy in Darwin's theory.

[x] 1802, R. Faulder, London.

[xi] *The Phenomenon of Man* at p.192, Diagram 4.

[xii] For online illustration of Conway's Game of Life simulation, go to: www.conwaylife.com

[xiii] *The Phenomenon of Man* at p.267.

[xiv] *Ibid. p.51, 66.*

[xv] *Ibid. at p.48.*

[xvi] *Ibid at p.163.*

[xvii] *Ibid. Man at p.180.*

[xviii] *Ibid. at p. 257.*

[xxi] Dawkins, Richard. (2006) *The blind watchmaker: why the evidence of evolution reveals a universe without design* New York: Norton.

[xxii] Ibid.

[xxiii] Darwin Letter to Asa Gray; May 22, 1860.

[xxiv] *Not by Chance* Dr. Lee Spetner; New York, The Judaica Press, 1997, 1998, at p.103).

[xxvii] Luskin, Casey; EN Evolution News, Biologist Michel Denton Revisits his Argument that Evolution is a 'Theory in Crisis"; 10.22.17.

[xxviii] Sears, Karen, "Development of Bat Flight; Morphology and Molecular Evolution of Bat Digits; Proceedings of the National Academy of Sciences of United States; 1p3; no 17 (2006).

[xxix] Ibid.

[xxxii] As recounted by Baretti, Giuseppe (1757), *The Italian Library Containing an Account of the Lives and Works of the Most Valuable Authors of Italy*, London, p.52.

[xxxiii] The Voyager spacecraft were launched in 1977 and are now 108 astronomical units from Earth. In April of 2015, the still functioning scientific instruments on Voyager 2 reported that the spacecraft had officially departed the outer edge of the solar system and entered the realm of interstellar space.

[xxxiv] Brain size is a good scientific indicator of intelligence and suggests that high level intelligence can also be found in other animals. Dolphin brain mass (1500-1,700 ml) exceeds human capacity (1300-1400 ml) and is four times the size of chimpanzees (400 ml). The study of brain structure suggests that the acoustic part of the dolphin brain is ten times the size of humans. Greater dependence on sound processing indicates a different kind of intelligence than homo sapiens. Cognitive skills in apes, chimpanzees, and dolphins demonstrate that knowledge can be transferred to individual members of these species. For example, skills acquired by dolphins in captivity (and released have been passed on to dolphins in the wild. Cognitive abilities have also been documented in elephants, whales, and octopuses.

[xxxv] As Darwin stated, "To suppose that the eye with all its inimitable contrivances for adjusting the focus to different distances, for admitting

different amounts of light, and for the correction of spherical and chromatic aberration, could have been formed by natural selection, seems, I freely confess, absurd in the highest degree". See, *Origin of the Species* (Charles Darwin; London: 2nd British ed; 1860).

xxxvi Interestingly, the primitive structures which feature eye spots known as planaria still exist today as does the simple pinhole structure of a nautilus eye. Such is direct evidence that the complexification process ensues notwithstanding the continued existence of more primitive forms which emerged as a result of evolution. As will be discussed later, it is the dual evidence of more primitive structures and subsequent complex iterations of same which points us in the direction of a higher power or God which likely coexists with Homo Sapiens here and now as we ourselves are in actuality a primitive structure which has been superseded by a biology and consciousness of greater complexity. We are simply an intermediate form.

xxxvii During the first 400,000 years after the Big Bang, no light was generated. Consequently, this primordial universe is invisible to astronomers using the most powerful instruments such as the Hubble Space Telescope. That changed on 9.14.15 when the LIGO experiment

(Laser Inferometer Gravitational Wave Observatory) at Caltech detected the world's first gravity waves. Predicted by Albert Einstein in 1916, gravity waves are not produced by light but disruptions caused in the fabric of spacetime which propagate in all directions at the speed of light. To better understand this phenomenon, think of spacetime as a great pool or water. Throw an object into the pool and ripples propagate in all directions. Now, substitute a massive explosion – the Big Bang – and imagine what ripples would have been created in the spacetime pool of the primordial universe. Theoretically, these ripples should be detectable by LIGO and would commence from the very inception of the Big Bang.

On 10.16.17, a second discovery was announced by scientists at MIT. LIGO detected gravitational waves originating from the collision of two neutron stars 130 million years ago. The emitted gravitational wave measured by LIGO lasted for over 100 seconds.

[xxxviii] Biologists were convinced that Coelacanths were extinct for millions of years until one was captured in 1938!

[xxxix] *The Phenomenon of Man* (Pierre Teilhard de Chardin; London: Collins; 1966; First French ed.; 1955).

[xl] There are numerous examples of other Emergent Border Phenomena, to wit, the potential existence of dark matter which physicists believe constitutes over 90% of matter in the universe. To date, there has not been scientific validation of this theory but astronomers believe it is the only possible explanation as to why objects in the universe are moving apart from each other at greater velocities contrary to our understanding of gravity. Another example is the theory of inflation which suggests that space itself expanded at speeds greater than the speed of light at the commencement of the so-called Big Bang. Once again, there is no scientific evidence to back this up. That said, the inexplicable observations aforesaid can best be described as being EBP's.

[xli] This observation recalls one of my favorite poems by Edgar Allan Poe herein included:

A Dream Within A Dream – by Edgar Allan Poe

Take this kiss upon the brow!
And, in parting from you now,

Thus, much let me avow-

You are not wrong, who deem,

That my days have been a dream;

Yet if hope has flown away

In a night or in a day,

In a vision or in none,

Is it therefore the less gone?

All that we see or seem

Is but a dream within a dream.

I stand amid the roar

Of a surf tormented shore,

And I hold within my hand

Grains of the golden sand-

How few! Yet how they creep

Through my fingers to the deep,

While I weep – while I weep!

O God! Can I not grasp

Them with a tighter clasp?

O God! Can I not save

One from the pitiless wave?

Is *all* that we see or seem

But a dream within a dream?

[xlii] I refer to this level of personal memory and consciousness as "Personal Morphic Field", borrowing a term from Rupert Sheldrake discussed in detail in Chapter 13.

[xliii] See Moorcroft, Fancis; *No. 5 Zeno's Paradox.*

[xliv] The variability of the expansion rate of space is evidenced by the observation that he rates of expansion of the universe is increasing over time. "Nobel Physics Prize Honors Accelerating Universe Find" (www.bbc.co/uk/news/science-environment-15165371) BBC News. 2011-10-04.

Additionally, the so-called inflationary epoch occurred at the negative 36 seconds after the Big Bang to 10 to the negative 33. However, it has *slowed* thereafter. "First Second of the Big Bang". How the Universe Works 3. 2014. Discovery Science.

[xlv] McDougall, 1927 at p.282.

[xlvi] See McDougall experimental data, 1938.

[xlvii] See, A New Science of Life: the hypothesis of formative causation, Los Angeles, CA; J.P. Tarcher, 1981; The Presence of the Past: morphic resonance at the habits of nature; New York, N.Y., Times Books, 1988.

[xlviii] For excellent discussion see, The Singularity is Near, Kurzweil, Raymond, Viking Books, 2005.

[xlix] C.G. Jung, The Archetypes and Collective Unconscious (London 1966) at p.43.

[l] Query the similarity between the inherited notion of the collective unconscious and Lamarckian theory of acquired characteristics.

[li] Jung, Collected Works, Vol. 9.1 (1959).

[lii] A personal example of the occurrence of a synchronicity arose when this writer had just completed this section of the book dealing with the existence of the Q Manifold and how a winning combination would instantaneously emerge within a quantum computer-generated operating system. After completion of this section, my two-year old daughter reached for a Rubik's Cube type puzzle which would require unlimited time to piece together a dinosaur versus an elephant. The timing of my daughter's selection of her next game was uncanny as it illustrated the operative effect of the Genesis 1.01 program on Earth resulting in the generation of life and intelligence.

[liii] Jung, Carl, Synchronicity, An Acausal Connecting Principle (First Princeton Bollingen ed, 1973).

[liv] Carl Jung recounts the following encounter with a patient: "My example concerns a young patient who, in spite of efforts made by both sides, proves to be psychologically inaccessible. The difficulty lay in the fact that she always knew better about everything. Her excellent education had provided

her with a weapon ideally suited to this purpose, namely a highly polished Cartesian rationalism with an impeccably geometrical idea of reality. After several fruitless attempts to sweeten her rationalism with a somewhat more human understanding, I had to confine myself to the hope that something unexpected or irrational would turn up, something that would burst the intellectual retort into which she had sealed herself. Well, I was sitting opposite her one day, with my back to the window listening to her rhetoric. She had an impressive dream the night before in which someone had given her a golden scarab – a costly piece of jewelry. She was still telling me this dream when I heard something behind me gently tapping on the window. I turned around and saw that it was a large flying insect that was knocking against the window pane from outside in the obvious effort to get into the dark room. This seemed to me very strange. I opened the window immediately and caught the insect in the air as it flew in. It was a scarabaei beetle, a common rose-chafer, who gold-green color most nearly resembles that of a golden scarab. I handed the beetle to my patient with the words, "Here is your scarab." The experience punctured a desired hole in her rationalism and broke the ice of her intellectual resistance. The treatment could now be continued with satisfactory results."

Ibid. at pp. 109-110.

lv Koestler, Arthur, The Roots of Coincidence ((Vantage, 1972) at p.87

lvi Tucker, Jim (University of Virginian Press) Jim Tucker, a professor of the University of Virginia, conducted interviews with children who had ironclad recollections of a past life. An example is a four-year-old boy named Ryan who recalled a prior life as an obscure Hollywood actor in the 1940's. He was inexplicably able to identify the names of stand in actors in old photographs and ultimately confirmed his own identity. The most incredible story concerns a two-year-old boy named James Leininger who loved toy airplanes. He was able to recall that he was involved in an airplane crash and the name of the boat he had flown off was the "Natoma". These observations were later confirmed as being historically accurate.

Another interesting area of exploration which suggests the existence of prior lives or field of information involves the possible connection between severe psychiatric disorders such as obsessive-compulsive disorder (OCD) and past life regression. Dr. Brian Weiss is a graduate of Columbia University and Yale Medical School and is the Chairman Emeritus Psychiatry at Mt. Sinai Medical Center in Miami. As a traditional psychotherapist he was skeptical when patients with severe disorders seemed to connect the traumas with past life events recalled in dreams or hypnosis. Dr. Weiss then studied these cases in detail with scientific rigor confirming that patient recollection of past life trauma through the process of past life regression provided an explanation for the occurrence of the problematic behavior. See, Many Lives, Many Masters: The True Story of a Prominent Psychiatrist, His Young Patient, and the Past-Life Therapy That Changed Both Their Lives (1988).

[lvii] Most people are unaware that a British mathematician, Charles Babbage (1791 to 1871), designed and partially constructed the first programable computer. This steam powered Goliath, the "Difference Engine" was never fully completed due to the abrasive personality of the inventor and lack of funding. For a musical rendition of this interesting story, go to the audio presentation of "A Different Engine" by this author available at www.prx.org (search for "A Different Engine").

[lviii] See, What is Artificial Intelligence (Copeland, Jack; www.alanturing.net; May 2000).

[lix] "Mind, Brains, and Programs; Behavioral and Brain Sciences, 1980.

[lx] The test was proposed by Alan Turing in his paper, "Computing Machinery and Intelligence"; Turing, 1950, at p. 460. The test would determine whether a human being could distinguish between having a conversation with a computer or another person, If the computer successfully passed itself off as a human participant, the Turing Test would be passed.

[lxi] Refer to discussion at Chapter 2 of this work by computer scientist Andrew Wade of a self-replicating e-organism named Gemini which appeared spontaneously at generation 33,699,586 of the Game of Life computer program.

[lxii] Consider, for example, the e-progression of Gemini over a billion generations. If self-replicating e-structures analogous to RNA were detected at generation 33,699,586, would it not be theoretically possible for complex e-organisms to evolve in subsequent generations akin to organic evolution. Query whether these organisms could develop e-central nervous systems and ultimately e-neurons as the generations continue to mount. The final result could exhibit the complexification observed in human neurons as the predicate of thought and consciousness.

[lxiii] Assume that the Game of Life e-ecosystem could be modified to accommodate a specific e-organism. For example, why not simulate an extinction event akin to the impact of the massive asteroid which hit the Yucatan Peninsula 65 Million Years ago and wiped out the dinosaurs. This extinction event led to the harsh environment which may have engendered the accelerated evolution of mammals and ultimately man. Why not replicate this event within the framework of the computer program and observe how the e-organism evolves?

[lxiv] For those who ponder theological/philosophical questions, the creation of intelligent e-life by humans is analogous to our biblical accounts of a higher power creating humans in God's image. Is it possible that the achievement of science and technology in the realm of artificial intelligence allows for the creation of sentient entities *by* sentient entities? If that is correct, then would this new e-world be encompassed within the boundaries of the artificial reality of AI itself? Although this proposition may seem incredulous, it was seriously proposed by University of Oxford philosopher, Nick Bostrum in 2003 suggesting that an advanced society would be capable of producing such an artificial world. Astronomer Neil

deGrasse Tyson believes that there is a 50/50 chance that all of us may be living in such a simulated world. The fact that science has uncovered universal mathematical rules which underpin reality is also suggestive that we may be living within the rigid rules of a simulated game. As stated by Max Tegmark, an MIT cosmologist, "If I were a character in a computer game, I would also discover eventually that the rules seemed completely rigid and mathematical." Of course, if such speculation proved correct, it would certainly raise serious questions as to where Teilhard's Arrow is pointing and the nature of the all-powerful sentient force at the apex of everything at Omega. I will let the reader fill in the blanks on this one! For an interesting discussion on the issue of simulated realities, go to *Scientific American*; 4.7.16; Are we living in a Computer Simulation? Moskowitz, Clara.

[lxv] Refer to my prior discussion on whether the hyper-consciousness of the Noosphere and Omega beyond would be dependent on the continued existence of biological life. The question would be the same.

[lxvi] Stephen Hawking warns that artificial intelligence could end mankind. "The development of full artificial intelligence could spell the end of the

human race... Humans, who are limited by slow biological evolution, couldn't compete and would be superseded. See, Rory Cellar-Jones, www.BBC.com ; 12,2,14. Elon Musk, CEO of SpaceX echoes the same concerns warning that artificial intelligence is "our biggest existential threat." Ibid.

lxvii See, www.motherboard.com; Koebler, Jason; 7.9.15. Researcher made an organic computer using four wired-together rat brains. The experiment was conducted at Duke University. The chief researcher, Miguel Nicolelis was able to utilize the "rat Brainnet" to perform "do useful computational problems such as classification, image processing, storage and retrieval of tactile information, and even weather forecasting. See also, "Building an Organic Computing Device with Multiple Interconnected Brain", www.nature.com; Science Reports; 7.9.15.

lxviii This discussion recalls my current experience raising our 19-month-old daughter. At a certain juncture, my wife and I observed Charley's sudden fascination with exploring her immediate environment with a focus on inanimate objects. As I write this end note, she is catapulting around a hotel lobby pausing every few seconds to ask – "DAT" (baby contraction for "what's that?". My wife and I answer, "a chair", "a lamp", etc. When Charley pauses in front of a full-length mirror it all becomes exceedingly interesting as she observes herself as a separate entity apart from the aforesaid hotel furniture. She seems to admire herself and smiles. Such is the beginning of self-awareness, no computer in the world is capable of this complex function. But let us assume that a complex bottom-up system is programmed in the foreseeable future to ask questions about its environment and interface with a human "trainer". I can conceive of some advanced iteration proceeding with "DAT?" and ending up far down the AI continuum at - "I think therefore I am".

[lxix] *A New Kind of Life* (Wolfram, Stephen: Wolfram; Wolfram Media, Inc. 2002).

[lxx] *Ibid. at p. 715.*

[lxxi] *Ibid. at p. 351-352.*

[lxxii] The same logic is utilized to predict that quantum computers will be able to decipher encrypted passwords almost simultaneously whereas the traditional binary system would take hundreds of years to arrive at the correct result following a linear path.

[lxxiii] *The Phenomenon of Man* (Pierre Teilhard de Chardin; London: Collins; 1966; First French ed.; 1955) at p.310.

[lxxiv] Evidence of human violence/warfare dates back over 13,000 years. The remains of systematic murder are found at an ancient cemetery dig in Northern Sudan and in Nataruk, Kenya where over 50% of skeletons have embedded warheads. There is also substantial evidence of primate violence relating to chimpanzees. Renowned researcher Jane Goodall observed the primate behavior of chimpanzees in Africa and observed that her initial view that animals were "rather nicer in behavior than humans" changed in 1974. She studies to groups of warring chimpanzee tribes which she later referred to as the "Gambee Chimpanzee War". Violent conflict between two groups of chimps in Gombe National Park in Tanzania had a profound effect on Goodall. She observed, "For several years I struggled to come to terms with this new knowledge. Often when I woke up in the night, horrific pictures sprang unhidden in my mind. Satan (one of the apes) cupping his hand below Sniff's (another Ape) chin to drink the blood that welled from a great wound to his face; Old Rodolf, unusually so benign, standing upright to hurl a four pound rock at Godi's prostate body; Jomeo tearing a strip of skin from De's thigh; Figan charging, hitting, again and again the stricken

quivering body of Goliath." Memoir, Through a Window: My Thirty Years with the Chimpanzees of Gombe; Goodall; 2010; pp. 128-29.

[lxxv] C G Jung to Miguel Serrano, Zurich, 14 September 1960; in M Serrano, *Jung & Hesse: A Record of Two Friendships* (New York: Schocken Books, 1968), pp. 84-85.

[lxxvi] This raises the interesting question as to whether the graphic and violent content of our media and entertainment products have the effect of accessing the stored memory in evil archetypes. The barrage of sexually explicit and violent content on cable television, for example, may have a measurable effect on the rise of violent behavior. In fact, numerous empirical studies have confirmed this causal connection between violent content and subsequent conduct.

[lxxvii] *The Phenomenon of Man* (Pierre Teilhard de Chardin; London: Collins; 1966; First French ed.; 1955) at p.310.

[lxxviii] Explaining Hitler, The Search for the Origins of His Evil (Rosenbaum, Ron, De Capo Press; 1998, 2014).

lxxix Ibid. at p.214.

lxxx Ibid. at p.215.

lxxxi *The Phenomenon of Man* (Pierre Teilhard de Chardin; London: Collins; 1966; First French ed.; 1955) at p.310.

lxxxii *The Science of Evil – on Empathy and the Origins of Cruelty* (Basic Books, 2011).

lxxxiii Ibid. at p. 6.

lxxxiv Ibid. at p.7. As Barin-Cohen acknowledges, this concept also goes back to Martin Buber's famous book, *I and Thou*.

lxxxv *The Phenomenon of Man* (Pierre Teilhard de Chardin; London: Collins; 1966; First French ed.; 1955) at p.310.

lxxxvi *Washington Post* (7.1.16) "How a scientist learned to work with exorcists".

lxxxviii A fascinating area known as Electronic Voice Phenomena (EVP) might shed light on the issue as to whether memory in some autonomous form survives death. EVP relates to voices which are inexplicably captured on recording devices which appear to be communications from individuals who have died. It was first reported by Konstantins Raudive who presented evidence of brief words or phrases which were recorded on devices placed in sound proof rooms. He made over 100,000 recordings which he described as communications with departed spirits. Evidence documents that these electronic "voices" are capable of responding to questions in real time suggesting that there is a conscious interaction involved. The author has met many of the principals in this field and was impressed with their qualifications and professionalism. Although the author cannot speak for the validity of all researchers in the EVP realm, no evidence of fraud or fabrication was uncovered.

An interesting speculation is whether the use of recording devices (digital and analog) collapse the waves of stored information in the noosphere and manifest as audio and visual messages/signals verified by EVP researchers.

For further information on this fascinating area, go to https://vimeo.com/115815330 which presents the documentary, The AFTERLIFE FILES, written and produced by filmmaker, Todd Moster.

www.ingramcontent.com/pod-product-compliance
Lightning Source LLC
Chambersburg PA
CBHW051306220526
45468CB00004B/1233